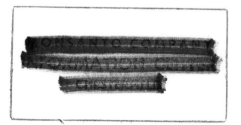

Molecular Biology of B Cell Developments

Cytokines

Vol. 3

Series Editor
C. Sorg, Münster

KARGER

Basel · München · Paris · London · New York · New Delhi · Bangkok · Singapore · Tokyo · Sydney

Molecular Biology of B Cell Developments

Volume Editor
C. Sorg, Münster

30 figures, 1 color plate and 6 tables, 1990

Basel · München · Paris · London · NewYork · New Delhi · Bangkok · Singapore · Tokyo · Sydney

Cytokines

Library of Congress Cataloging-in-Publication Data
Molecular biology of B cell developments / volume editor, C. Sorg.
p. cm. — (Cytokines; vol. 3)
Includes bibliographical references.
Includes index.
1. B Cells – Differentiation – Molecular aspects. 2. Cytokines – Physiological effects.
3. Growth factors – Physiological effect.
I. Sorg, Clemens. II. Series.
[DNLM: 1. B-Lymphocytes – immunology. 2. Lymphocyte Transformation]
ISBN 3–8055–5191–6

Drug Dosage
The authors and the publisher have exerted every effort to ensure that drug selection and dosage set forth in this text are in accord with current recommendations and practice at the time of publication. However, in view of ongoing research, changes in government regulations, and the constant flow of information relating to drug therapy and drug reactions, the reader is urged to check the package insert for each drug for any change in indications and dosage and for added warnings and precautions. This is particularly important when the recommended agent is a new and/or infrequently employed drug.

© Copyright 1990 by S. Karger AG, P.O. Box, CH–4009 Basel (Switzerland)
ISBN 3–8055–5191–6

Contents

Preface

One question that still dominates basic immunological research is where and how do B lymphocytes develop from stem cell origins. Because of its fundamental importance it is an area of great research efforts. The expectations run high as a number of consequences for the clinic can be envisaged such as the control of immunoglobulin class switch in immediate-type hypersensitivity or the down-regulation of specific antibody production in certain autoimmune diseases.

As we know, B cell development is driven by antigen and by numerous cell-cell interactions, notably with epithelial cells, dendritic cells, macrophages and T lymphocytes. These events, which involve a cellular and molecular biology of outmost complexity, are step by step unveiled and today we are faced with a bewildering wealth of questions, approaches and achievements.

As certain cytokines are more and more recognized to exert key functions in B cell development, it was felt timely to invite a group of reputed authors to review major aspects of this field. The topics covered range from the micro-environment of B cell development over immunoglobulin class switching to the *scid* mouse mutant.

I wish to thank the authors and I hope this volume serves the purpose well in providing some guidance for a complex but exciting theme of immunology.

Clemens Sorg

Sorg C (ed): Molecular Biology of B Cell Developments.
Cytokines. Basel, Karger, 1990, vol 3, pp 1–23

Role of Anti-IgM Antibodies, Bacterial Lipopolysaccharide and T Lymphokines in Mouse B Cell Activation: Analysis of Cell Cycle Response and Immunoglobulin RNA Processing

Una Chen

Basel Institute for Immunology, Basel, Switzerland

The development of immunoglobulin M (IgM)-secreting plasma cells from resting B cells involves three major steps: (1) activation as seen by increasing cell size and Ia antigen expression following recognition of the antigen by its immunoglobulin receptor; (2) proliferation of these activated B cells (lymphoblasts), and (3) differentiation of B lymphoblasts into immunoglobulin-secreting plasma cells. The recognition of antigen by antigen receptors can be mimicked by cross-linking them with anti-antigen receptor antibodies (anti-μ is commonly used) [1, 2].

We have demonstrated that cross-linking B cell antigen receptors with anti-μ antibodies recruits the small resting B cells from the G_0/G_1 phase into the cell cycle, but these cycling cells become arrested in the G_1 phase of their second cell cycle [3]. Further proliferation and differentiation require help from T cells and accessory cells and/or their lymphokines. We characterized the roles of T lymphokines in potentiating the B cell cycle progression after anti-μ treatment. In addition, we investigated the expression of immunoglobulin genes accompanying these cellular events and compared these results with those obtained by polyclonal activation of B cells with bacterial lipopolysaccharide (LPS).

LPS bypasses the requirements for antigen recognition, thus leading to coupling of proliferation and differentiation [4–6]. The cellular activation provoked by LPS is accompanied by 3 major molecular changes: (1) transcriptional activation of immunoglobulin genes (both heavy and light chains); (2) RNA processing of the immunoglobulin genes, especially the 3'-end cleavage and polyadenylation, and (3) transcriptional activation of the IgM J chain gene [7–10]. The molecular events controlling these 3 steps are regulated positively in LPS-treated B cells.

In contrast to the multiple action of LPS, exposure to anti-μ causes B cells to enter the replication cycle but has little effect on immunoglobulin gene transcription (step 1). Further proliferation and differentiation events require external stimuli [11, 12]. Steps (2) and (3) can be influenced positively and negatively by the addition of lymphokines. Using a combination of cell kinetic and molecular approaches, we attempted to clarify the role of anti-μ and T lymphokines in the B cell cycle and in inducing differentiation (immunoglobulin gene expression). We present evidence that: (1) interleukin-4 (IL-4) and interleukin-5 (IL-5) are co-competence factors of anti-μ, and IL-4 alone is an S phase progression factor; (2) recombinant-interferon (r-IFN) is not a competence factor either alone or in combination with anti-μ or LPS; rather, it is an S-phase-potentiating factor; (3) transcriptional enhancement of immunoglobulin gene loci is an important event in the activation of B cells, and (4) exposure of B cells to anti-μ alone and/or to r-IFN does not induce immunoglobulin RNA processing.

Materials and Methods

Tissue Culture Conditions, Stimuli and Lymphokines

C57BL/6 nu/nu mice were obtained from Bomholgaart (Denmark) and housed at the Basel Institute for Immunology. The isolation of resting B cells, tissue culture conditions and the 5-bromo-2'-deoxyuridine (BrdU)/Hoechst 33258 dye (Hoechst)/ethidium bromide (EB) bivariate flow-cytometric method have been described [3, 13, 14]. Resting mouse B cells were cultured either in medium alone or with different stimuli for a total of 4 days. T lymphokines were included in the culture together with anti-μ either at seeding or 1 day later. r-IFN and concanavalin-A-stimulated T cell supernatant (CASUP) were included in the culture 36 h after seeding. The cells were harvested at successive time points as indicated in the Results (see legends and figures). For transcriptional run-on assay and RNA isolation, the activated lymphoblasts were separated from the unstimulated and dead cells by the Percoll gradient centrifugation method (Pharmacia, Uppsala, Sweden) prior to cell lysis.

LPS of 20 μg/ml final concentration was from *Salmonella typhimurium* (Sigma, St. Louis, Mo., USA). Affinity-purified goat anti-mouse μ-chain F(ab')$_2$ fragments (anti-μ) of 10 μg/ml final concentration were received from Jackson Labs (West Grove, Pa., USA). Ionomycin (ION; 0.1 μM) and phorbol dibutyrate (PDBu; 10 ng/ml final concentration) were purchased from Calbiochem (Luzern, Switzerland). Rat r-IFN (10 U/ml final concentration) was a gift of P. v.d. Meide [15]. Mouse CASUP (5–10%, v/v, final concentration) contains IL-2, IL-3, IL-4, IL-5, r-IFN and T-cell-replacing factor [16].

Recombinant mouse IL-2, IL-3, IL-4 and IL-5, and human IL-6 were supernatants from X63Ag8.653 plasmacytoma cells which had been transformed with BPV vectors containing cDNAs of the corresponding IL [17] and were gifts of F. Melchers. They were used at 30 U/ml final concentration in culture; this titer was optimal (data not shown). Biochemically purified T-cell-replacing factor (BCDF) devoid of any other known IL was processed according to Müller et al. [16] and was depleted of r-IFN by anti-r-IFN affinity column at a final concentration of 5–10%. Since there is no standard unit of this lymphokine, the concentration was titrated to the minimal amount required for a positive activity (data not included).

Transcriptional Run-On Assay

The in vitro transcriptional run-on assay was performed as described previously [13, 18], with the following modification: B cells in a freshly isolated resting state, or after anti-μ stimulation, anti-μ plus r-IFN or CASUP treatment, and LPS stimulation were harvested, centrifuged and subjected to lysis buffer containing Triton X-100. The mixtures were centrifuged through a cushion of sucrose buffer. Pellets (which contain mainly nuclei) were harvested and subjected to RNase treatment. The nuclei were washed again and stored at $-70\,°C$ until ready for run-on assays. 10^8 nuclei/reaction were used together with 250–500 μCi ^{32}P-UTP in the reaction mixture for 15 min in the presence of RNase inhibitor (Promega, Switzerland).

The in vitro elongated and labeled RNA transcripts were isolated, aliquoted and hybridized for 4 days at $42\,°C$ with Southern blots which contained DNA fragments coding for immunoglobulin Cμ, IghE, δ, γ, κ, actin, H-2, rRNA and phage marker-containing mouse repetitive DNA elements. The total amount of ^{32}P-UTP-labeled RNA was estimated by trichloroacetic acid precipitation; the difference between 3 experiments was less than 3-fold. After hybridization, blots were washed, RNase-treated and exposed at $-70\,°C$ with X-films and an intensity screen. The resulting signals were scanned with a densitometer, normalized and calculated. Only the transcriptional rate at the linear range was used for comparison.

RNA Blotting

RNA isolation from different cultured B cells, gel electrophoresis, isolation, transfer, hybridization and ^{32}P nick translation of DNA probes were performed as described [13], except that nylon-base Gene Screen (Amersham, Buckinghamshire, UK) was used. Following electrophoresis, the agarose gels were treated briefly with alkali and then neutralized before blotting.

Results

B Cell Cycle

B Cells Are Cycling in Response to Anti-μ and T Lymphokines. We recently introduced a flow-cytometric assay for studying mouse B cell proliferation [3, 14]. We apply this method to assay the response of resting

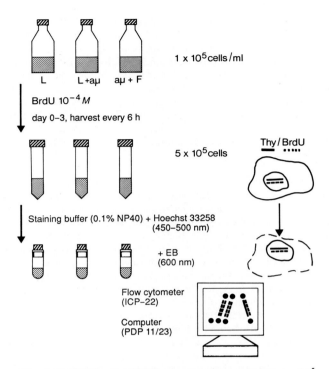

Fig. 1. BrdU/Hoechst-EB flow-cytometric method. 1×10^5 mouse B cell/ml are cultured with different stimuli [LPS, plus anti-μ and anti-μ plus T lymphokines (F)] in the presence of 10^{-4} M BrdU for 3 days in flasks. The cells are harvested every 6 h. Aliquots of cells (5×10^5 cells) are transferred to centrifuge tubes and centrifuged. The cells in pellets are collected and frozen at −20 °C in freezing medium. After thawing the cells are stained with staining buffer containing 0.1% NP40 and Hoechst for 15 min, and then with EB for another 15 min. The detergent NP40 increases the permeability of the cell membrane and allows the penetration of dyes into the nucleus. The dyes stain DNAs in the nucleus at their native chromatin configuration. The stained cells were measured in an ICP-22 flow cytometer at wavelengths of 450–500 nm for Hoechst and above 600 nm for EB. The data were analyzed in a PDP 11/23 computer with a software program designed by Peter Rabinovitch. The two-dimensional cytogram shown on the TV screen reveals up to 4 distinct cell cycles. The x-axis is a linear channel of Hoechst fluorescence, and the y-axis is a linear channel of EB fluorescence.

B cells to LPS, anti-μ and T lymphokine stimulation. As shown in figure 1, cells were grown in the continuous presence of BrdU and harvested from 24 to 72 h at 6-hour intervals. Unstimulated and stimulated cells were analyzed by their differential staining pattern to the Hoechst and EB fluorescent dyes.

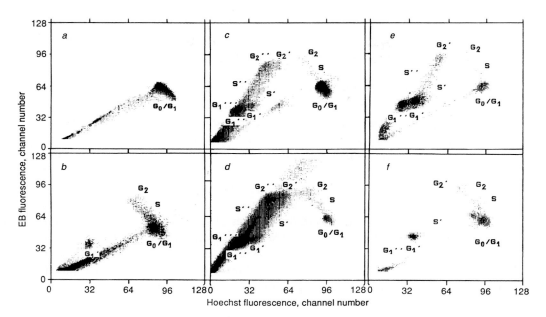

Fig. 2. BrdU/Hoechst-EB bivariate cytogram of B cells upon induction. Resting mouse B cells were analyzed after culturing with different stimuli. The cells were cultured in the presence of BrdU for 72 h, harvested and subjected to analysis as described previously [3, 14]. *a* Unstimulated B cells after culturing in medium alone. *b* Anti-μ stimulation. *c* LPS stimulation. *d* LPS plus anti-μ stimulation. *e* Anti-μ plus IL-4 stimulation. *f* Anti-μ plus IL-5 stimulation. The x- and y-axes are linear channels from 1 to 128.

The theories and detailed description of the dye-binding property of this assay have been published elsewhere [14, 19]. Here, I briefly explain the fluorescence signals in the cytogram and then propose a new model of the B cell cycle. The advantages of this assay include the following: (1) the continuous progress of up to 4 cell cycles can be visualized; (2) the detailed cell cycle compartmental distribution of cells can be calculated accurately, and (3) the minute difference of the cell behavior in the cycle can be detected.

Unstimulated Cultured B Cells Remain Noncycling. The cytogram shown in figure 2 illustrates that B cell cultures consist of identical clusters of nonreplicating cells (G_0/G_1 cell fraction) at 72 h of incubation. A subpopulation of these cultured resting B cells undergoes massive changes in their dye-binding properties. These cells are represented by a trail of fluorescent signals that extends from the region of the G_0/G_1 cluster to the origin of the cytogram; such major changes in dye binding are likely to reflect nuclear pyknosis and impending cell death [Kubbies et al., pers. commun.].

Anti-µ Recruits Resting B Cells into the Cell Cycle. Exposure of B cells to anti-µ leads to nonstoichiometric dye binding (nuclear pyknosis) in a considerable portion of these cells. This is shown in the cytogram of figure 2b by the prominent signal trail extending from the cluster of noncycling cells to the origin. However, a small portion of the G_0/G_1 cells enter the S phase. This cycling fraction grows, and 22.3% of the intact cell population has entered the cell cycle by 60–72 h. BrdU incorporation is represented in this cytogram by increasing EB, but decreasing Hoechst axis fluorescence. The G_2/M phase nuclei are seen as the uppermost portion of the first-cycle S phase signal track. At the end of culture, 27.6% of activated cells remain in the S and G_2/M phase of the first cycle. 5.0% of the responding B cells have become daughter cells in the second-cycle G_1 compartment (designated G_1'), which appears as a cohort of signals positioned to the left of the first-cycle signals. The postmitotic cells accumulated in the G_1 compartments of the second cycle with very few cells entering the S and G_2 compartments of the second cycle or beyond. The phenomenon of apparent arrest in the G_1 phase of the second cell cycle is specific for anti-µ stimulation, since we have not observed a similar pattern with other mitogens.

LPS Is an Effective Mitogen of B Cell Growth. As shown in figure 2c, LPS stimulates approximately the same proportion of resting B cells to enter the cell cycle as exposure to anti-µ alone (23.9% at 60 h) [see also ref. 3 for detailed calculation]. However, in contrast to anti-µ, these LPS-treated cells progress through 4 cycles by 72 h of stimulation. In the cytogram, the second-cycle G_1' cluster extends upward toward the right as the S′ and G_2'/M fluorescence signals. The fluorescence signals of this second S phase run opposite to those of the first-cycle S phase. This mirror image is caused by a switch from uni- to bifilarly substituted DNA during the second round of DNA replication in the presence of the halogenated nucleoside analogue BrdU [see also ref. 14 and 19 for detailed explanation]. The replicating cells in the third cell cycle are the signal track to the left of those in the second cell cycle. The G_1, S and G_2/M phases of the third cycle (designated G_1'', S″ and G_2'') are separated from those of the second cycle by a small shift to the left. To the left and overlapping with G_1'' are the flourescence signals of G_1''' of the fourth cycle. The cells have properly migrated through 4 consecutive cell

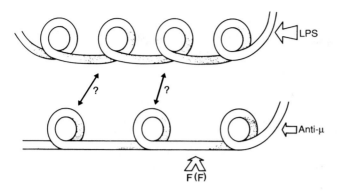

Fig. 3. Model of cell-cell interaction within B cell subpopulations. In response to LPS stimulation, a subset of resting B cells progresses up to 4 consequent cell cycles. Upon anti-μ stimulation, another subset of B cells reaches only 1–2 cycles and peculiarly arrests at G_1 compartments. The further progression of the cell cycle relies on T lymphokines [F(F)] or LPS co-stimulation. The combined stimulation of B cells with LPS and anti-μ activates the majority of B cells into the cell cycle synergistically. This suggests strongly that activated B cells might secrete factors which participate in recruiting more resting B cells into the cell cycle.

cycles until they slowly face degradation (evidenced by signals exhibiting less than the regular G_1' or G_1'' fluorescence). LPS is, therefore, a much more effective cell cycle progression factor than anti-μ (and T lymphokines; see following section). The average cell-doubling time from 30 to 72 h has been calculated as being 11–12 h [see ref. 3 for calculation]. However, there is a tiny population of B cells cycling rapidly in response to LPS alone or LPS plus anti-μ with an estimated doubling time of 6 h [3; unpubl. data]. These data are consistent with those estimated by Zhang et al. [20] in B cells from germinal centers, by Askonas and North [21] in splenic B cells and by Meyer et al. [22] in pre-B cell lines.

Cell Interact within B Cell Subpopulations. There is cell-cell interaction between anti-μ-responding and LPS-responding B cell subpopulations. As shown in figures 2b–d and 3 (the model), 17–30% of B cells are LPS-responsive and 20–30% are anti-μ-responsive. In the presence of both LPS and anti-μ for 3 days, over 95% of B cells are out of the resting stage. The data suggest that LPS and anti-μ stimulate resting B cells synergistically and that there are possible factors secreted by activated B cells which are involved in the recruiting of resting B cells into replication cycles. The stimulating/

regulating B cell lymphokines have also been suggested from the studies of del Guercio et al. [23], Gordon [24], Kiely et al. [25] and Braun et al. [26]. One of these B cell lymphokines have been cloned as IL-10 [M. Howard, pers. commun.].

T Lymphokines Augment B Cell Activation (by Another Activator). There are at least 5 cloned and well-characterized T lymphokines which have been shown to be active in regulating B cell activation [27–29]. These are IL-2, IL-4, IL-5, IL-6 and r-IFN. Depending on the study model, there are conflicting opinions regarding which factor is active and what are its defined roles in B cell replication and/or differentiation (see below, Molecular Analysis, pp 9–10). For example, IL-2 and IL-4 are thought to have counter-effects on B cell activation, either on primary mouse B cells [30, 31], on B cell lines [e.g. BCL-1; Koshland, pers. commun.] or on human B cells [32]. In the systems where IL-4 acts as a positive co-stimulator of antireceptor antibodies, IL-2 appears to be inactive [unpubl. data], while other data [30, 31, 33] demonstrated that IL-2 plays an inducing role in replication and differentiation. In our system, IL-4 is a co-competence factor of antireceptor antibodies. IL-4 alone has also been shown to be an S phase progression factor. The combined stimulation of resting B cells by anti-μ and IL-4 can activate a majority of cells progressing to the G_1 phase of the third cycle (fig. 2e). It is peculiar that B cells activated with anti-μ-plus IL-4 were preferentially distributed in the G_1 and S phases of the second and third cycles; few cells accumulate in G_2. This distribution is reminiscent of results obtained with embryonic fibroblasts. Interestingly, long-term B cell cultures can be established when anti-μ- plus IL-4-treated B cells are infected with immunoglobulin promoter and enhancer driving *myc* and *ras* oncogenes containing retrovirus (RIM virus) [Chen and Marcu, unpubl. observ.]. This observation suggests that the enrichment of B cells at the G_1 compartment might provide receptors for retrovirus infection and subsequently continuous replication of B cells.

The role of IL-5 in B cell activation is again debatable. It appears to be a good proliferation and differentiation lymphokine in total splenic B cells or activated B cell subpopulations, or the Ly-1$^+$ AJ9 cell line [unpubl. data and ref. 29, 30, 34], but weak or nonactive in a single-cell culture system [35] or in true resting B cell populations [unpubl. data]. However, it has been shown to be highly active in peritoneal Ly-1$^+$ B cell populations [35].

Like IL-4, IL-5 has been shown to be a co-competence factor of antireceptor antibodies on small B cells. The cell cycle progresses to the G_1' phase of the second cycle. Very few cells advance to the third cycle (fig. 2f). Moreover, with activated B cells IL-5 is also an S phase progression factor.

r-IFN Maintains Dual Roles in B Cell Activation. r-IFN plays dual roles in mouse B cell proliferation. It has a cell-cycle-progressing potential on well-activated B cells. On the other hand, it inhibits the mitogenic properties of other activators [36; unpubl. data] and alone cannot stimulate resting B cells to replication.

The in vitro activation of resting B cells to the r-IFN-responsive stage can be achieved by antireceptor antibodies and/or LPS. The timing of activation is critical. The first 0–30 h are nonresponsive. From 36 h on, more B cells enter and progress through the cell cycle. The S phase of the second and third cycles is more pronounced in the presence of r-IFN. Clearly, r-IFN acts as a factor promoting B cell growth [3].

However, it is puzzling that r-IFN has a negative effect on B cell replication. r-IFN is not a competence factor either alone or together with other activators. Moreover, it inhibits the mitogenic properties of such activators as LPS, anti-μ and T lymphokines (data not shown) [36]. If r-IFN is included in the early stage of activation, the pattern of inhibition is peculiar. There are more cells in the G_1 phase of the second/third cycle. The G_0/G_1 fluorescence signals tend to migrate to less fluorescence and toward the origin of the 'nuclear decay' track.

A New Model of the B Cell Cycle. Based on our data, we propose a new model of B cell activation upon anti-μ and lymphokine stimulation. As shown in figure 4, small B cells can be activated to enter the cell cycle and progress through 3 or 4 (LPS; not shown) cycles depending on the property of activators and B cell subpopulations. The true virgin B cells (designated as B_0 subpopulation) are less responsive, while the more activated (and possible in vivo exposure to antigen/mitogen) B cells (designated as B_1 subpopulation) are fully responsive to stimulators. Although these 2 subpopulations share similar amounts of heat-stable protein (J-11D), they are different by the existence of IL-5-responding immunoglobulin-secreting B cells (only in the B_1 subpopulation) and by their reactivity to anti-μ stimulation (1 or 2 cycles) and lymphokines (as differential compartmental promoting factors). Anti-μ can exit resting B cells into 1 cycle (B_0, fig. 4a) or 2 cycles (B_1, fig. 4b). Anti-μ and IL-4/IL-5 can exit resting B cells from the G_0/G_1 phase into the cell cycle and complete 3 cycles (B_0, fig. 4c, and B_1, fig. 4d), while the tested lymphokines (especially IL-4 and r-IFN; fig. 4d) act as cell cycle progression factors.

We could not observe any mitogen that acts on the G_2 compartments [37, 38]. It is, however, possible to accumulate B cells in specific cell cycle compartments by treatment of B_0 and B_1 cells with anti-μ, which causes the G_1 arrest of the second or third cell cycle depending on the cell subpopulation and by anti-μ plus IL-4, which can enrich cells at the G_1 and S compartments. The method might provide the possibility of enriching B cells in specific cell cycle compartments for further molecular analysis.

Molecular Analysis of Immunoglobulin RNA Transcription upon B Cell Activation

The molecular events that occur after activation of primary B cells with LPS alone and LPS plus anti-μ have been studied by several groups [8, 10, 13, 39–43].

LPS Enhances Immunoglobulin Transcription and Causes μm-μs Processing. The data from several laboratories have consistently supported that the transcriptional activation of B cells by LPS alone is an important event, and the μ-3'-end processing is an equally important posttranscriptional event. There is evidence for both reduction of RNA polymerase II density [39], and cleavage and polyadenylation [10, 44]. This conclusion is based on transcriptional run-on assays and Northern blotting on the μ-δ locus. In our system of transcriptional run-on assay, δ-region is consistently loaded with significant amounts of RNA polymerases, while in Weiss et al.'s system [39] there is a drastic decrease in RNA polymerase activity in this region when B cells are activated. We could also detect unspliced long μ-δ RNA transcripts when primary B cells were stimulated under certain conditions, for example with LPS plus anti-μ plus r-IFN [40; unpubl. data]. Upon the activation of B cells there is a significant increase in immunoglobulin transcription. The mechanism of $\mu m \rightarrow \mu s$ switch [45–47] has been debated. Although mRNA of the μm form appears developmentally earlier than the μs form, μm polyadenylation is a more effective process than μs polyadenylation in vivo [46, 47] as well as in vitro [44]. The increasing concentration of μ-RNA transcripts/substrates upon B cell maturation seems to favor the polyadenylation event at the weaker μs site. There are at least 3 distinct protein factors involved in the polyadenylation event. The question of what makes the μm polyadenylation stronger than μs polyadenylation needs to be answered.

In addition to immunoglobulin gene activation, the transcriptional activation of IgM J chain gene is an important event in secreting IgM pentamer [7, 9].

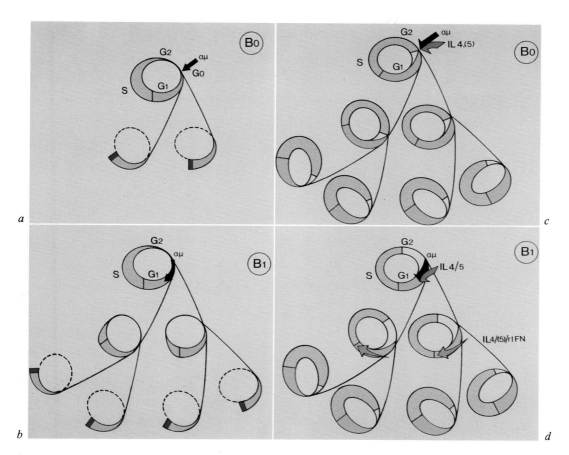

Fig. 4. (For legend see reverse side.)

Fig. 4. Model of mouse B cell cycle after anti-μ and lymphokine stimulation. Small B cells are separated into two populations based on the density: population B_0 (density = 1.085 g/ml) and population B_1 (density = 1.077 g/ml). Upon the activation with anti-μ (aμ) alone or anti-μ plus lymphokines, the cells exit from the G_0/G_1 phase. In the presence of anti-μ alone, small B cells have completed the first cycle and arrested at the G_1' compartment (population B_0, *a*) or completed 2 cycles and arrested at the G_1'' compartment (population B_1, *b*). There are very few cells distributed at G_2 phases in both populations. mIL-4 acts as a co-stimulator of anti-μ, exits the majority of B cells from the G_0/G_1 phase, and progresses through 3 cycles (*b, c*). mIL-4 alone is also a cell-cycle-progressing factor (*d*). mIL-5 has some effect on resting B cells (population B_0, *c*). However, its effect is mainly on activated B cells (population B_1, *d*) as a co-competence factor of anti-μ as well as a progressing factor. r-IFN alone or in combination with other stimuli has no effect in exiting resting B cells; it acts as an S phase progression factor on activated B cells (*d*). Yellow represents the G_1 compartment, pink represents the S compartment, white represents the G_2 compartment, and red represents G_1 arrest. The position of colors in the cell cycle indicates the percentage of cells in the different compartments. Anti-μ activation is shown by the black arrows, IL-4/5 as co-competence factors by the green arrows, and IL-4/(5)/r-IFN as S phase progression factor by the blue arrows.

LPS plus Anti-μ Enhance Immunoglobulin Transcription, but Interfere with μm-μs Processing. When primary mouse B cells are activated with LPS and anti-μ antibodies, anti-μ exerts a dominant negative effect in inhibiting the normal differentiation of B cells to plasma cells by LPS alone.

Molecular studies have shown that posttranscriptional events are the primary target of anti-μ inhibition.

The 3'-end μ-RNA processing, i.e. the switch of μm to μs, shows interference. In the presence of r-IFN, we can detect multiple unprocessed μ-RNA transcripts [40]. We postulate that small nuclear ribonucleoproteins/ enzymes controlling RNA polyadenylation are affected.

In order to understand this inhibitory phenomenon, we must understand the biochemical events of μm and μs polyadenylation and the mechanism which controls this switch process. A nuclear extract system was established from stimulated primary B cells and tumor cells. This in vitro system enabled us to test the hypothesis [Chen and Virtanen, unpubl. data].

Transcription of Immunoglobulin Loci Occurs during B Cell Activation by Anti-μ plus r-IFN. In B cell activation, cell proliferation is required for differentiation. The molecular events that occur when B cells are stimulated with anti-μ plus crude natural T lymphokines have been reported [12]. The roles of anti-μ and recombinant T lymphokines, especially r-IFN, in B cell differentiation have also been characterized. In contrast to other treatments that activate B cells, such as LPS or anti-μ and CASUP, treatment of resting B cells with anti-μ plus IL-4 or r-IFN does not yield secreted IgM. The combined treatment with anti-μ plus IL-5 of a very small population of B cells yields IgM secretion. It was shown that r-IFN alone does not activate resting B cells, and the combination of anti-μ and r-IFN does not stimulate B cells to immunoglobulin secretion (data not shown). In order to assess the molecular events asssociated with cell proliferation, we decided to analyze the intracellular immunoglobulin gene expression during B cell activation upon stimulation with anti-μ plus r-IFN. In addition, the immunoglobulin expression of B cells stimulated with anti-μ alone or in combination with CASUP was compared to that of B cells stimulated with LPS.

The transcription rates were measured by in vitro transcriptional run-on assays (fig. 5; summarized in table 1) for B cells in the resting stage (fig. 5a), or stimulated with anti-μ (fig. 5b), anti-μ plus r-IFN (fig. 5c), anti-μ plus CASUP (fig. 5d) and LPS (fig. 5e). Resting B cells yield weak transcriptional signals for μ, δ and κ. Marker (lane 1), which contains cloned mouse repetitive DNA elements [48], and 28 S r-RNA (lane 9) served as standards.

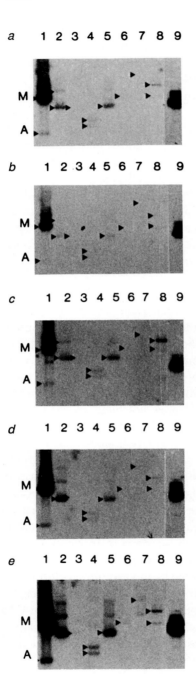

Fig. 5. In vitro transcriptional activity of B cells upon stimulation. Freshly isolated primary B cells, designated B_0 *(a)*, or cells stimulated with different stimuli [anti-μ *(b)*, anti-μ plus r-IFN *(c)*, anti-μ plus CASUP *(d)* and LPS *(e)*] were harvested on day 3. Nuclei were isolated, and in vitro transcription assays were performed. Filters containing recombinant DNA fragments were incubated with ^{32}P-labeled RNA for 4 days at 42 °C. One representative result of these assays is shown. Lane 1 = Marker (M) and actin (A); lane 2 = Cμ; lane 3 = γ3; lane 4 = IghE; lane 5 = κ; lane 6 = H-2; lane 7 = δ secretory; lane 8 = δ membrane; lane 9 = rRNA. Autoradiograms were scanned with a densitometer; the surface areas of the relevant fragments were divided by the length of hybridizable sequences and by the number of nuclei, and then plotted against RNA input. The values are defined as the transcriptional rates and are given in table 1. The calulation of both marker and rRNA is not included in table 1, since they were not transcribed by polymerase II, and served only as the markers to standardize the signals.

Table 1. In vitro transcriptional rate of immunoglobulin loci upon B cell activation by different stimuli

Gene locus	Nuclei from resting B cells	Nuclei from cells treated with			
		anti-μ	anti-μ + r-IFN	anti-μ + CASUP	LPS
Cμ	1.42	1.19	26.5	33.4	62.4
IghE	0.9	0.72	10.2	20.3	38.8
Cκ	1.9	0.9	20.1	30.7	49.4
Actin	0.8	0.9	10.8	20.4	36.9

The transcriptional rates of nuclei from B cells stimulated with anti-μ plus r-IFN and anti-μ plus CASUP are similar (fig. 5c, d; table 1). These rates are approximately 20-fold higher than those of nuclei exposed to anti-μ alone (fig. 5b) and of resting B cells (fig. 5a), and about one half of those of LPS-treated nuclei (fig. 5e). The transcriptional rate of LPS-treated nuclei is at least 30- to 40-fold higher than that of resting B cells.

Transcription of the actin gene is also increased significantly upon activation. This is not surprising since actin transcription is required for cell proliferation [10]. As was previously found in the LPS system [8, 10], RNA polymerase II activity in the δ-region is still detectable after B cell activation (fig. 5a–e, lanes 7 and 8), although no transcription is detected in the γ-region (fig. 5a–e, lane 3). These data suggest that, in the primary B cell system, the 3'-end of μ-mRNA is determined mainly by cleavage and polyadenylation, not by transcriptional termination [39].

Anti-μ and r-IFN Interfere with RNA Processing. The analysis of μ-chain RNA from B cells stimulated with anti-μ, anti-μ plus lymphokines or LPS is shown in figure 6a. Resting B cells (lane 1) give little mature μ-mRNA, which is exclusively of the membrane form. One day after anti-μ stimulation (lane 2), there is no detectable mature μ-RNA or μ-RNA transcript. Two days later (lane 3), a μ-RNA transcript (designated Pμ$_3$) of about 6.5 kilobases (kb) is visible. Three days later (lane 4) the amount of Pμ$_3$ increases, but mature μ-mRNA is still not detectable.

Upon addition of r-IFN (lane 5), 3 additional unspliced μ-RNA transcripts appear besides the μ-RNA transcript (Pμ$_3$) already visible upon anti-μ treatment. These RNA transcripts are designated Pμ$_1$, Pμ$_{2a}$ and Pμ$_{2b}$, and estimated to be 18, 12.0 and 10.8 kb, respectively. When quantified by densitometry, these 3 transcripts represent about 60% of the amount of RNA produced by LPS-treated B cells. In all of these RNA preparations (lanes 2–5) there is no mature μ-mRNA. Mature μ-mRNA appears only upon the addition of multiple lymphokines (CASUP; lane 6). The amount of mRNA detected in anti-μ plus CASUP-treated B cells is comparable to that from B cells stimulated with LPS alone (lane 7). The IgM J chain is expressed only after anti-μ plus CASUP and LPS treatments (data not shown). The mature μ-mRNAs shown in lanes 6 and 7 represent the secretory forms.

Negative Effect of Anti-μ Cannot Be Replaced by PDBu plus ION. It appears that the negative effect of anti-μ on the splicing of the μ-RNA primary transcript can be overcome by the addition of multiple lymphokines. I have asked whether this negative effect could be mimicked by the combined treatment with PDBu plus ION. These reagents, which activate protein kinase C, have been shown to mimic anti-μ in activating resting B cells to replication [49–51]. In our hands, these reagents potentiate the LPS effect in activating B cells into the cell cycle, but they do not mimic the action of anti-μ alone (data not shown). Figure 6b shows the results of replacing anti-μ with PDBu plus ION. PDBu plus ION treatment does not interfere with μ-RNA processing as anti-μ does (lanes 2–4). However, in the presence of r-IFN (lane 5), there is an accumulation of multiple μ-RNA transcripts of 18, 12, 10.8 and 6.5 kb. Work is in progress to further characterize the transcripts shown in figures 6a and b. In addition, a small quantity of mature μ-mRNA is detected. These data suggest that the synergy of anti-μ and r-IFN (fig. 6a, lane 5) exerts a more profoundly negative effect on μ-RNA processing than does the combination of PDBu plus ION plus r-IFN (fig. 6b, lane 5).

Discussion

Mouse B Cells Have a Unique Cell Cycle Clock
As shown in the model (fig. 3, 4), the transition of mouse B cells into cell cycle compartments requires a cascade of external stimuli. Depending on the activation pathways, B cells can either continuously progress in the cycles or discontinuously arrest at different cycle compartments. The former example is shown in the activation of B cells by LPS stimulation; the latter event is represented by the stimulation of B cells with anti-μ and lymphokines. Our

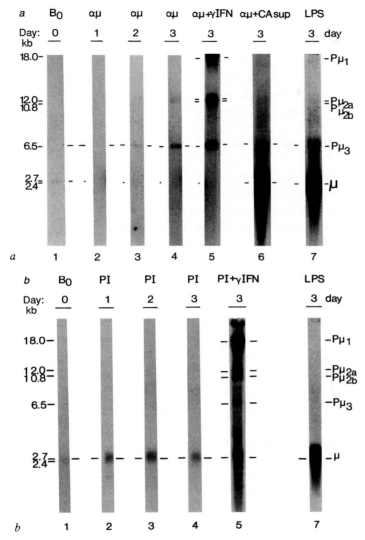

Fig. 6. RNA analysis of B cells after induction. The analysis was performed with the total RNA from resting B cells (lane 1, day 0; *a, b*), and from B cells stimulated with different stimuli, harvested on day 1 (lane 2), day 2 (lane 3) and day 3 (lanes 4–7). RNA from an equal amount of cultured cells was extracted, electrophoresed through a 1% agarose gel and transferred to filters. Hybridization probes were nick-translated and hybridized to RNA blots. *a* Stimulation with anti-μ (lanes 2–4), anti-μ plus r-IFN (lane 5), anti-μ plus CASUP (lane 6) and LPS (lane 7). *b* Replacement of anti-μ with PDBu plus ION (PI). Lanes 2–4 = PI; lane 5 = PI plus r-IFN; lane 7 = LPS.

results, therefore, suggest that mouse B cells have a unique cell cycle clock. The restriction points are at the transitions from the G_0 to the G_1 phase and from the G_1 to the S phase compartment. The first point is demonstrated most prominently by anti-µ stimulation, and the second is demonstrated by r-IFN and IL-4 (and also IL-5) supplementation. The ability of B cells to cycle depends on the strength and type of stimuli. For example, LPS stimulation causes cells to migrate through 4 consecutive cycles within 72 h. Upon anti-µ stimulation the cellular response is limited to 1 cycle, while anti-µ plus r-IFN yields transition through 2, and anti-µ plus IL-4/IL-5 yields transition through 3 cycles.

Immunoglobulin Transcriptional Enhancement
Occurs upon B Cell Maturation

In the primary cell system, exposure of resting B cells to only anti-µ does not yield significant increases in the rate of immunoglobulin transcription. Transcriptional enhancement occurs after supplementation by r-IFN or CASUP of anti-µ-treated B cells, which indicates that B cell proliferation is not a sufficient criterion for active transcription of the immunoglobulin loci. This event requires additional help by T lymphokines or LPS [10, 13, 40].

Results reported with cell lines representing different stages of the B cell ontogeny have suggested that the rate of transcription does not change significantly from the pre-B cell stage through the mature B cell stage up to the plasma cell stage [52, 53]. These results are in sharp contrast to those obtained in the primary cell system employed in this report and those of earlier studies of LPS-activated B cells [8, 10], where strong transcriptional enhancement was found in response to appropriate activation. However, this contradiction may be spurious, since permanent cell lines are, by definition, permanently activated. Hence, experiments with such lines compare activated cells at one stage (e.g. B cells) with activated cells at another stage (e.g. pre-B cells). Experiments with primary cells, as reported herein, compare resting B cells with activated B cells.

T Lymphokines Are Required for Induction of Mature µ-mRNA

The analysis of µ-mRNA processing during B cell activation reveals that the formation of µm RNA in B cells activated via the anti-µ pathway requires additional stimuli (lymphokines other than r-IFN) to overcome their defective splicing. Activation via LPS stimulation circumvents this requirement. There are several possible explanations for the splicing defect caused by anti-µ and r-IFN stimulation. It could represent either a deficiency or an active suppression of the splicing process. Since the addition of anti-µ to LPS [13] and the addition of anti-µ plus r-IFN to LPS-activated B cells [40] caused a similar defect, it is likely that an active suppression of splicing occurs during exposure to these two reagents.

We have further investigated the lymphokines in CASUP that might provide the regulatory stimulus. None of the recombinant factors (i.e. IL-2 to IL-6), alone or in combination, can perform this effect. Therefore one candidate remains: T-cell-replacing factor [54, 55], which has yet to be cloned. This lymphokine, originally studied and defined at the cellular level, is a late-acting differentiation factor which converts activated nonsecreting lymphoblasts into immunoglobulin-secreting plasma cells.

The molecular events in the transition from activated lymphoblasts to plasma cells involve processing of the 3'-end and splicing of the immunoglobulin primary transcript, as well as IgM J chain activation [7]. The observation that the J chain is not expressed whenever there is a block of µ-RNA processing (as shown in this and earlier studies [13, 43]) suggests that there might be a coordinate regulation of µ-RNA processing and J chain activation. The study of Kubagawa et al. [56] suggests that they are separate events.

It remains to be elucidated whether these events are regulated by another undefined lymphokine or a combination of growth and differentiation factors present in the CASUP. Because only extremely small amounts of T-cell-replacing factor can be purified from CASUP, it was not possible to isolate sufficient quantities to perform the appropriate molecular studies for testing this hypothesis.

Anti-µ Exerts Multiple Effects on B Cell Activation

Because PDBu plus ION treatment could not mimic anti-µ in blocking immunoglobulin RNA precursor processing and the combination of anti-µ plus r-IFN causes an accumulation of immunoglobulin transcripts, immunoglobulin RNA splicing is not a direct consequence of B cell activation or transcription. Anti-µ must provide both the positive (which activates resting B cells to replicate) and the negative signal (which arrests the activated B cells at G_1 compartments and which also interferes with RNA splicing and seems to be executed via another, as yet undefined, pathway). By studying the early events of B cell mitogenesis, Mond et al. [57] have observed that stimulation of B cells by anti-IgD may involve pathways other than the protein-kinase-C-dependent mechanism. Perhaps the receptor itself (either intact or a partial product), a µ-like protein, or µ-chain-binding proteins [58, 59] bind to small nuclear ribonucleoproteins and interfere with RNA splicing.

Interferon-γ Has a Regulatory Role in Immunoglobulin
RNA Processing

IFN elevates the level of several proteins and induces new gene expression in fibroblasts [60]. Among the proteins induced by IFN are $2',5'$-(adenosine)$_n$ [$2',5'$-(A)$_n$] synthetases, which convert ATP to $2',5'$-(A)$_n$ and pyrophosphate [61, 62]. The $2',5'$-(A)$_n$ intermediates serve as substrates to activate RNase L, which cleaves single-stranded RNA [63, 64]. $2',5'$-(A)$_n$ might also control the activities of proteins other than RNase L [64]. $2',5'$-(A)$_n$ synthetases have been implicated in the processing of heterogeneous nuclear RNA (hnRNA) [65].

r-IFN treatment of B cells might increase the level of $2',5'(A)_n$ synthetases, which would increase the level of endogenous RNases and thereby enhance the rate of the immunoglobulin RNA splicing process. However, this prediction is in conflict with our experimental observations, i.e. exposure to r-IFN inhibits the RNA-processing machinery and leads to the accumulation of unprocessed RNA transcripts.

We have studied $2',5'$-(A)$_n$ synthetase activity in r-IFN-treated B cells. No increase in the level of $2',5'$-(A)$_n$ synthetase mRNA was detected with a DNA probe coding for $2',5'$-(A)$_n$ synthetase [66] (data not shown). Moreover, using the rRNA assay, we found no increase in the activity of $2',5'$-(A)$_n$-dependent endo-RNase [67] (data not shown).

These observations suggest that the activation of IFN-responding elements which cause the inhibition of immunoglobulin hnRNA processing must occur via another, as yet undefined, mechanism. The defect of splicing seems to be specific for immunoglobulin genes; the expression of unrelated genes such as H-2 and actin is not affected (data not shown).

Primary B cells activated by anti-μ and exposed to such partial T cell help as r-IFN might serve as important models for the understanding of B cell regulation. The differential mechanisms of B cell activation may be clarified by the following observations: B cells are activated into the nonimmunoglobulin-secreting blast cell stage; immunoglobulin RNAs are transcribed but persist in the long unspliced precursor forms, and the IgM J chain mRNA is not expressed. We postulate these cells to be either abortive B cells or memory precursor cells.

Adoptive transfer experiments [unpubl. data] of these activated and nonsecreting B cells have suggested that they might be the memory precursor B cells. This mechanism may, therefore, serve to expand rapidly the pool of memory (precursor) cells without the burden of concomitant IgM production.

Acknowledgments

I thank J. Chebath, M. Ravel, T. Wirth and those referred to in the paper of Chen-Bettecken et al. [13] for plasmids, R. Silverman for endo-RNase assay and P. v.d. Meide for r-IFN. I also thank C. Baker, G. Wetzel, W. Schaffner, C. Steinberg and H. Höhn for critical reading and help with this manuscript and U. Kaempf, M. Kuhn, H. Seyschab and R. Friedl for excellent technical assistance and flow-cytometric measurements. I especially thank H. Höhn for introducing me to the BrdU/Hoechst-EB two-dimensional flow-cytometric assay to study the mouse B cell cycle and for helping me form the new theoretical model of B cell activation. The Basel Institute for Immunology was founded and is supported by F. Hoffmann-La Roche & Co., Ltd., Basel, Switzerland.

References

1   Tony HP, Schimpl A: Stimulation of murine B cells with anti-Ig antibodies: Dominance of a negative signal mediated by the Fc receptor. Eur J Immunol 1980; 10:726–729.
2   Parker DC: Separable helper factors support B cell proliferation and maturation to Ig secretion. J Immunol 1982;129:469–474.
3   Seyschab H, Friedl R, Schindler D, et al: The effects of bacterial lipopolysaccharide anti-receptor antibodies and recombinant interferon on mouse B cell cycle progression using 5-bromo-2′-deoxyuridine/Hoechst 33258 dye flow cytometry. Eur J Immunol 1989;19:1605–1612.
4   Andersson J, Sjoeberg O, Moeller G: Induction of immunoglobulin and antibody synthesis in vitro by lipopolysaccharide. Eur J Immunol 1972;2:349–353.
5   Melchers F, Andersson J: Synthesis, surface and secretion of immunoglobulin in bone marrow derived lymphocytes before or after mitogenic stimulation. Transplant Rev 1973;14:76–130.
6   Coutinho A, Probor G, Petterson S, et al: T cell-dependent B cell activation. Immunol Rev 1984;78:211–224.
7   Koshland ME: Presidential address: Molecular aspects of B cell differentiation. J Immunol 1983;131:i–iv.
8   Yuan D, Tucker PW: Transcriptional regulation of the μ–δ heavy chain locus in normal B lymphocytes. J Exp Med 1984;160:564–583.
9   Lamson G, Koshland ME: Changes in J-chain and μ-chain RNA expression as a function of B-cell differentiation. J Exp Med 1984;160:877–892.
10 Chen-Bettecken U, Wecker E, Schimpl A: Transcriptional control of μ- and κ-gene expression in resting and bacterial lipopolysaccharide-activated normal B-cells. Immunobiology 1987;174:162–176.
11 Howard M, Paul WE: Regulation of B-cell growth and differentiation by soluble factors. Annu Rev Immunol 1983;1:307–333.
12 Nakanishi K, Cohen DI, Blackman M, et al: Ig RNA expression on normal B cells stimulated with anti-IgM antibody and T-cell-derived growth and differentiation factors. J Exp Med 1984;160:1736–1751.
13 Chen-Bettecken U, Wecker E, Schimpl A: IgM RNA switch from membrane to

secretory form is prevented by adding anti-receptor antibody to bacterial lipopoly-saccharide-stimulated murine primary B-cell cultures. Proc Natl Acad Sci USA 1985; 82:7384–7388.

14 Poot M, Schmitt, H, Seyschab H, et al: Continuous bromodeoxyuridine labeling and bivariate ethidium bromide/Hoechst flow cytometry in cell kinetics. Cytometry 1989;10:222–226.

15 van der Meide PH, Dubbeld M, Vijverberg K, et al: The purification and characterization of rat gamma interferon by use of two monoclonal antibodies. J Gen Virol 1986;67:1059–1071.

16 Müller G, Hübner L, Schimpl A, et al: Partial characterization and purification of murine T cell-replacing factor, TRF-I: Purification procedures and gel electrophoretic analysis. Immunochemistry 1978;15:27–32.

17 Karasuyama H, Melchers F: Establishment of mouse cell lines which constitutively secrete large quantities of IL-2, 3, 4 or 5 using modified cDNA expression vectors. Eur J Immunol 1988;18:97–104.

18 Mather EL, Nelson KJ, Haimovich J, et al: Mode of regulation of immunoglobulin μ- and δ-chain expression varies during B-lymphocyte maturation. Cell 1984;36:329–338.

19 Rabinovitch PS, Kubbies M, Chen YC, et al: BrdU/Hoechst flow cytometry: A unique tool for quantitative cell cycle analysis. Exp Cell Res 1988;174:309–318.

20 Zhang J, MacLennan ICM, Liu YJ, et al: Is rapid proliferation in B centroblasts linked to somatic mutation in memory B cells? Immunol Lett 1988;18:297–300.

21 Askonas BA, North JR: The lifestyle of B cells: Cellular proliferation and the invariancy of IgG; in Watson JD (ed): Cold Spring Harb Symp Quant Biol. Cold Spring Harbor, Cold Spring Harbor Laboratory, 1976, vol XLI, pp 749–759.

22 Meyer J, Jäck HM, Ellis N, et al: High rate of somatic point mutation in vitro in and near the variable-region segment of an immunoglobulin heavy chain gene. Proc Natl Acad Sci USA 1986;83:6950–6953.

23 del Guercio P, del Guercio MF, Katz DH: Characterization of murine interleukin B by a monoclonal antibody. Nature 1987;324:445–447.

24 Gordon J: Mechanisms of B lineage growth and differentiation – points of decision and of possible transformation: Growth and differentiation factors; in Melchers F, Potter M (eds): Mechanisms of B Cell Neoplasia, 1987. Basel, Editiones Roche, 1987, pp 75–86.

25 Kiely JM, Braun J, Unanue ER: Autoregulation of B cell growth by immunoglobulin M autoantibody. J Immunol 1985;135:1040–1045.

26 Braun J, Citri Y, Baltimore D, et al: B-Lyl1 cells: Immortal LY-1+ B lymphocyte cell lines spontaneously arising in murine splenic cultures. Immunol Rev 1986;93:5–15.

27 Swain SL, McKenzie DT, Dutton RW, et al: The role of IL-4 and IL-5: Characterization of a distinct helper T cell subset that makes IL-4 and IL-5 (Th₂) and requires priming before induction of lymphokine secretion. Immunol Rev 1988;102:77–105.

28 Sideras P, Noma T, Honjo T: Structure and function of interleukin 4 and 5. Immunol Rev 1988;102:189–21.

29 Karasuyama H, Rolink A, Melchers F: Recombinant interleukin 2 or 5, but not 3 or 4, induces maturation of resting mouse B lymphocytes and propagates proliferation of activated B cell blasts. J Exp Med 1988;167:1377–1390.

30   Zubler RH, Lowenthal JW, Erard F, et al: Activated B cells express receptors for, and proliferate in response to, pure interleukin 1. J Exp Med 1984;160:1170–1183.

31   Llorente L, Crevon MC, Karray S, et al: Interleukin (IL) 4 counteracts the helper effect of IL 2 on antigen-activated human B cells. Eur J Immunol 1989;19:765–769.

32   Wetzel GD: Interleukin 5 regulation of peritoneal Ly-1 B lymphocyte proliferation, differentiation and autoantibody secretion. Eur J Immunol 1989;19:1701–1707.

33   Blackman MA, Tigges MA, Minie ME, et al: A model system for peptide hormone action in differentiation: Interleukin 2 induces a B lymphoma to transcribe the J chain gene. Cell 1986;47:609–617.

34   Lernhardt W, Karasuyama H, Rolink A, et al: Control of the cell cycle of murine B lymphocytes: The nature of α- and β-B-cell growth factors and of B-cell maturation factors. Immunol Rev 1987;99:241–262.

35   Braun J, Krall WJ, Clark ME, et al: Inducible Ig heavy chain switching in an IgM⁺ Ly-1 B cell line: Evidence for a state of switch commitment. J Mol Cell Immunol 1988;4:105–119.

36   Rabin EM, Mond JJ, Ohara J, et al: Interferon-γ inhibits the action of B cell stimulatory factor (BSF)-1 on resting B cells. J Immunol 1986;137:1573–1576.

37   Melchers F, Andersson J: B cell activation: Three steps and their variations. Cell 1984;37:715–720.

38   Melchers F, Lernhardt W: Three restriction points in the cell cycle of activated murine B lymphocytes. Proc Natl Acad Sci USA 1985;82:7681–7685.

39   Weiss EA, Michael A, Yuan D: Role of transcriptional termination in the regulation of μ-mRNA expression in B lymphocytes. J Immunol 1989;143:1046–1052.

40   Chen U: Anti-IgM antibodies inhibit IgM expression in lipopolysaccharide-stimulated normal murine B-cells: Study of RNA metabolism and translation. Gene 1988;72:209–217.

41   Högbom E, Martensson EL, Leanderson T: Regulation of immunoglobulin transcription rates and mRNA processing in proliferating normal B lymphocytes by activators of protein kinase C. Proc Natl Acad Sci USA 1987;84:9135–9139.

42   Flahart RE, Lawton AR: Mechanism of suppression of lipopolysaccharide-driven B cell differentiation by anti-μ antibodies: Evidence for a *trans*-acting repressor of transcription. J Exp Med 1988;17:865–873.

43   Virtanen A, Sharp PA: Processing at immunoglobulin polyadenylation sites in lymphoid cell extracts. EMBO J 1988;7:1421–1429.

44   Peterson ML, Perry RP: The regulated production of μ_m and μ_s is dependent on the relative efficiencies of μ_s poly(A) site usage and the Cμ4-to-M1 splice. Mol Cell Biol 1989;9:726–738.

45   Galli G, Guise JW, McDevitt MA, et al: Relative position and strengths of poly(A) sites as well as transcription termination are critical to membrane versus secreted μ-chain expression during B-cell development. Gene Dev 1987;1:471–481.

46   Galli G, Guise J, Tucker PW, et al: Poly(A) site choice rather than splice site choice governs the regulated production of IgM heavy-chain RNAs. Proc. Natl Acad Sci USA 1988;85:2439–2443.

47   DeFranco AL, Gold MR, Jakway JP: B-lymphocyte signal transduction in response to anti-immunoglobulin and bacterial lipopolysaccharide. Immunol Rev 1987;95:161–176.

48   Bijsterbosch MK, Meade CJ, Turner GA, et al: B lymphocyte receptors and polyphosphoinositide degradation. Cell 1985;41:999–1006.

49 Klaus GGB, O'Garra A, Bijsterbosch MK, et al: Activation and proliferation in mouse B cells: VIII. Induction of DNA synthesis in B cells by a combination of calcium ionophores and phorbol myristate acetate. Eur J Immunol 1986;16:92–97.

50 Gerster T, Picard D, Schaffner W: During B-cell differentiation both enhancer activity and transcription rate of immunoglobulin heavy chain genes are high before mRNA accumulation. Cell 1986;45:45–52.

51 Jäck HM, Wabl M: Immunoglobulin mRNA stability varies during B lymphocyte differentiation. EMBO J 1988;7:1041–1046.

52 Dutton RW, Falkoff R, Hirst M, et al: Is there evidence for a non-antigen specific diffusable chemical mediator from the thymus-derived cell in the initiation of the immune response?; in Amos F (ed): Progress in Immunology. New York, Academic Press, 1971, pp 355–365.

53 Schimpl A, Wecker E: Replacement of a T-cell function by a T-cell product. Nature 1972;237:15–16.

54 Kubagawa H, Burrows PD, Grossi CE, et al: Precursor B cells transformed by Epstein-Barr virus undergo sterile plasma-cell differentiation: J-chain expression without immunoglobulin. Proc Natl Acad Sci USA 1988;85:875–879.

55 Mond JJ, Feuerstein N, Finkelman FD, et al: B-lymphocyte activation mediated by anti-immunoglobulin antibody in the absence of protein kinase C. Proc Natl Acad Sci USA 1987;84:8588–8592.

56 Haas IG, Wabl M: Immunoglobulin heavy chain binding protein. Nature 1983;306:387–389.

57 Bole DG, Hendershot LM, Kearney JF: Posttranslational association of immuno-globulin heavy chain binding protein with nascent heavy chains in nonsecreting and secreting hybridomas. J Cell Biol 1986;102:1558–1566.

58 Langyel P: Biochemistry of interferons and their actions. Annu Rev Biochem 1982;51:251–282.

59 Hovanessian AG, Brown RE, Martin EM, et al: Enzymic synthesis, purification, and fractionation of (2′,5′)-oligoadenylic acid. Meth Enzymol 1981;79:184–193.

60 Samanta H, Dougherty JP, Lengyel P: Synthesis of $(2′,5′)(A)_n$ from ATP: Character-istics of the reaction catalyzed by $(2′,5′)(A)_n$ synthetase purified from mouse Ehrlich ascites tumor cells treated with interferon. Nature 1981;289:414–420.

61 Wreschner DH, McCauley JW, Skehel JJ, et al: Interferon action: Sequence specific-ity of the $ppp(A2′p)_nA$-dependent ribonuclease. EMBO J 1987;6:1273–1280.

62 St. Laurent G, Yoshie O, Floyd-Smith G, et al: Interferon action: Two $(2′,5′)(A)_n$ synthetases specified by distinct mRNAs in Ehrlich ascites tumor cells treated with interferon. Cell 1983;33:95–102.

63 Nilsen TW, Maroney PA, Robertson HD, et al: Heterogeneous nuclear RNA promotes synthesis of (2′,5′)oligoadenylate and is cleaved by the (2′,5′)oligoadenyl-ate-activated endoribonuclease. Mol Cell Biol 1982;2:154–160.

64 Hovanessian AG, Laurent AG, Chebath J, et al: Identification of 69-kd and 100-kd forms of 2-5A synthetase in interferon-treated human cells by specific monoclonal antibodies. EMBO J 1987;6:1273–1280.

65 Silverman RH, Krause D: Analysis of anti-viral mechanisms: Interferon-regulated 2′5′-oligoadenylate and protein kinase systems; in Clemens MJ, Morris AG, Gearing AJH (eds): Lymphokines and Interferons: A Practical Approach. Oxford, IRL Press, 1987, pp 149–193.

66 Wirth T, Glöggler K, Baumruker T, et al: Family of middle repetitive DNA sequences in the mouse genome with structural features of solitary retroviral long terminal repeats. Proc Natl Acad Sci USA 1983;80:3327–3330.

67 Chen-Bettecken U, Wecker E, Schimple A: Transcriptional and post-transcriptional control of Ig-gene expression in murine B-cells activated by LPS and anti-receptor antibodies; in Ferrarini M, Pernis B (eds): The Molecular Basis of B-Cell Differentiation. New York, Plenum Press, 1986, pp 29–32.

Una Chen, PhD, Basel Institute for Immunology,
Grenzacherstrasse 487, CH–4005 Basel (Switzerland)

Sorg C (ed): Molecular Biology of B Cell Developments.
Cytokines. Basel, Karger, 1990, vol 3, pp 24–60

B Cell Microenvironments during Antigen Stimulation

*Ernst Heinen, Cécile Kinet-Denoël, Alain Bosseloir,
Nadine Cormann, Léon J. Simar*[1]

Institute of Human Histology, University of Liège, Belgium

B cells as well as T cells undergo during their life two major maturation phases. The first one occurs in the absence of antigens (lymphopoiesis or lymphogenesis) in the fetal liver or bone marrow as regards the B cells and in the thymus as concerns the T cells. The second one develops in the presence of antigen in the peripheral lymphoid organs (immunopoiesis or immuno-genesis) [1]. Since another chapter of the present volume describes the lymphopoiesis, we will focus our attention only on the immunopoiesis. This process starts with virgin B cells (mature B cells) which are produced by the bone marrow and express membrane IgM, IgD and other molecules enabling their recirculation or their contacts with other cells.

The encounter with antigens and the subsequent events do not occur in stochastic ways. An antigen first appears at peculiar sites. During or after the contact with the antigen, B cells must receive additional signals from surrounding cells to either undergo a differentiation into antibody-forming cells (AFCs; plasma cells) or to proliferate and then differentiate into memory cells or AFCs (fig. 1). The lymphoid cells concerned have received a wide variety of designations, resulting in many confusions. Our terminology will be the following: a lymphocyte is a small and quiescent cell (6–9 μm in diameter) with dense chromatin; an activated lymphoid cell (a blast cell) is a

[1] We are grateful to the 'Fonds de la Recherche Scientifique Médicale' and the Faculty of Medicine of Liège for their financial help, and thank Mrs. P. Dubois and C. Martin for typing the manuscript.

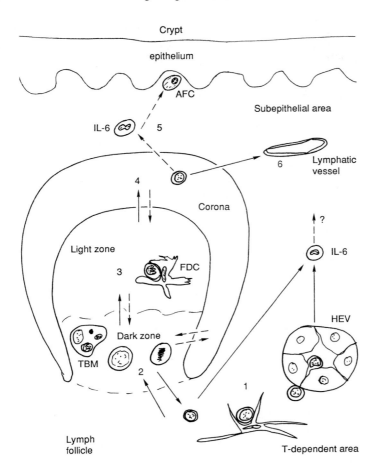

Fig. 1. Possible migration pathways of B cells in a human tonsil. B cells, after passing through HEVs, come in contact with T cells and antigen-presenting cells (1). When activated they gain the dark zone (2) and transform into centroblasts, in vicinity of tingible body macrophages (TBM) and FDCs. Centroblasts either mature to memory cells in the light zone (3) or emigrate through the interfollicular area (1) where they meet IL-6-secreting cells and are induced to transform into AFCs. Centrocytes (3) can transform into centroblasts or, more probably, small lymphocytes which leave the follicles through the corona (4), encounter IL-6 in the subcapsular area (5) and can transform into plasma cells beneath the epithelium. Memory B cells gain the efferent lymphatics (6) and recirculate. ⟶ = Probable route; --⟶ = hypothetical route.

large cell (10–20 µm) with dispersed chromatin and prominent nucleoli. The term lymphoblast designates a proliferating virgin cell in the bone marrow or thymus (lymphopoiesis); an immunoblast denotes an antigen-stimulated lymphoid cell in the follicular or thymus (T)-dependent zones. A centrocyte and a centroblast designate cells occurring, respectively, in the light and dark zones of germinal centers (GCs). Centrocytes are medium-sized cells with indented nuclei; centroblasts are large, basophilic cells with round nuclei and developed nucleoli. Plasmablasts are precursors of plasma cells (AFCs).

A novel encounter with the same antigen will drive the organism to produce a stronger response (higher immunoglobulin production, other immunoglobulin classes, higher affinities) than after the first contact; this is called the secondary response. It is, however, worth recalling that during the primary response all elements allowing the secondary response are formed, namely the memory cells. Owing to the progresses in immunology during the last years, it is now possible to propose characteristic features defining the memory B cells: clonal amplification; increase in immunoglobulin affinity [2]; isotype switch [3]; change of recirculation molecules (vascular addressins [4]), and loss of Fc receptors [5]. B cell clones are short-lived when unstimulated [6], but otherwise they have a long lifespan [7, 8].

Except for the clonal amplification, memory T cells which participate in the humoral response do not, apparently, exhibit the same features as memory B cells. They are phenotypically different from naive T cells (CD45; integrins), produce greater amounts of cytokines [interferon-γ and interleukin-4 (IL-4)] and also express higher-affinity receptors [6, 9, 10]. Memory T cells express increased levels of adhesion molecules (CD2, LFA-1; LFA-3; CD44; CDw29, ICAM-1); this suggests that facilitated cell interactions may account for at least some of the features of secondary immune responses [11].

The following sections will describe particular aspects of the humoral response: microenvironments for B cell activation, proliferation, maturation and terminal differentiation into AFCs. However, since lymphoid structures are frequently the subject of some confusion, it appears necessary to first review their histological definitions.

Histology of Lymphoid Tissues and Organs

A tissue is composed of cells specialized in order to express one or more of a physiological property [12]. Usually, four basic tissue types are considered: epithelial tissue; connective tissue; nervous tissue, and muscular tissue. The lymphoid cells are part of the connective tissue which is composed of cells, fibers and ground substance. Lymphocytes can simply infiltrate a connective tissue or assemble in lymphoid tissues or organs.

The bone marrow and thymus are the primary or central lymphoid organs since the lymphopoiesis occurs therein. The spleen, lymph nodes, tonsils, Peyer's patches and appendix are the secondary or peripheral lymphoid organs because they are the sites of the immunopoiesis. This distinction is, however, not absolute: the bone marrow also plays an important role in the immunoglobulin production and, thus, has a link with immunopoiesis; the ileal Peyer's patches are apparently also sites of lymphopoiesis [13]. Lymphoid cells can preferentially home in to mucous areas (gut, respiratory or genitourinary tracts) and infiltrate there tissues or assemble in follicles or T-dependent zones. This allows the definition of a peculiar histological entity: the mucous-associated lymphoid tissue (MALT) which will be described in detail below.

B or T lymphocytes are thus not tissues (as frequently stated) but cells composing tissues or organs.

Definition of a Microenvironment

The term 'microenvironment' is widely used but with different acceptations. For example, in one case it means erroneously the environment of the lymphoid cells in the whole lymph node or, in other cases, it is restricted to the immediate area around a single cell. Since the vicinity of a cell can be analyzed at the ultrastructural level and since most of the bioactive molecules of the immune system act like paracrine hormones at short distances, we will only designate the close vicinity of the cell considered or of a defined cell group with the term microenvironment. From the histological point of view, one can thus progress from the cell to its microenvironment, to the tissue, to the organ and then to the system or apparatus. For example, one can consider a lymphoid cell, placed in the microenvironment of an M cell in an epithelial dome of a Peyer's patch (lymphoid organ which is integrated in the MALT system).

A microenvironment is built up by cells (stromal, parenchymal, migrating cells), by their products (e.g. fibers, ground substance, bioactive molecules) and by the intercellular fluids (serum, lymph), and depends on physical features (temperature, pressure) and chemical constituents (e.g. ions, pH, concentration in O_2 and CO_2). Consequently, when studying a microenvironment, the histological structures composing it as well as its physiological status at a given time must be considered.

A cell can be in contact either with a single cell (in the case of emperipolesis) or with several cells; for example, a T cell can contact simultaneously a B cell, a macrophage and a follicular dendritic cell (FDC) [200]. Thus, a microenvironment around a cell can be homogeneous and stable or, on the contrary, complex and variable.

Architecture of Lymphoid Tissues

Various lymphoid tissues compose the lymphoid organs or are integrated in other tissues. An overview of the main lymphoid tissues is presented here, i.e. the lymph follicles, T-dependent areas and some peculiar zones as the medullary and Billroth cords, the marginal zone and the local infiltrations.

Lymph Follicles. The link between lymph follicles (or nodules) and the humoral immune response has been established long ago [14]. Unstimulated follicles (primary follicles) are spheric structures where mainly small B cells are found inside a reticular meshwork; few T cells, macrophages and FDCs are also present [15, 16]. The B cells are mostly bone-marrow-derived virgin cells (IgM$^+$, IgD$^+$, low MHC class II$^+$) and are able to recirculate [17]. The follicles are found inside the peripheral lymphoid organs, in the MALT, and can develop de novo in various sites of the organism after repeated antigenic stimulation.

When activated after antigenic stimulation, the follicles (secondary follicles) enlarge as a result of the development of the GCs. Their general shape is that of nodules (1–2 mm in diameter) but then they develop tails composed of emigrating cells gaining the medullary or Billroth cords. Their periphery (mantle zone or corona) contains a cell population of primary follicles and can have the form of a crescent (tonsil; fig. 1).

The GCs undergo cyclic changes. Few days after the induction with antigen, blast B cells (centroblasts) appear and divide. The typical sub-division in a dark zone (centroblasts) and a light zone (centrocytes) of the GCs is seen after 1–3 weeks and persists for a few weeks to several months. The centroblasts disappear in the end phase [18, 19]. During the first part of the primary response, most B cells (66%) int he GCs are IgM-positive and 10% are IgG-positive; when the antigenic stimulation persists or is repeated these proportions are reversed: 30% are IgM-positive and 72% are IgG-positive [20]. In tonsils and lymph nodes, most of these GC cells bear IgG; in Peyer's patches they produce mainly IgA [21, 22].

Medullary and Billroth Cords. Medullary cords can be considered as appendices of the lymph node follicles [12] but also as entities by their own

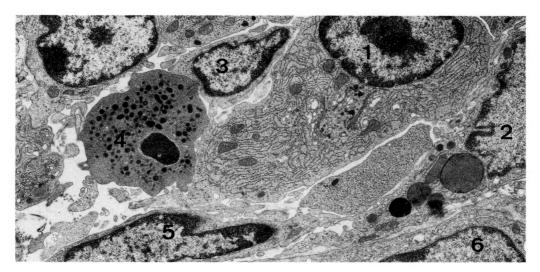

Fig. 2. Plasma cell (1) in a medullary cord of a mouse lymph node. It contacts simultaneously a macrophage (2), a small lymphocyte (3), a granulocyte (4) and a fibroblast (5) near the endothelium (6) of a sinus. ×7,900.

since they lodge their proper cell population. They are anchored in the paracortex; in fact, they are organized around blood vessels (arterioles and venules) and are connected to the tails of the secondary follicles. Cell types composing these cords are B cells (immunoblasts, plasmablasts, plasma cells, lymphocytes), T cells (recirculating lymphocytes), macrophages, reticular cells and polynuclear cells (fig. 2). Most of these cells migrate toward the medullary zone, namely the B cells, and then differentiate into AFCs. These plasma cells are retained by the endothelium of the medullary sinuses and secrete immunoglobulins which leave the lymph nodes by the efferent lymph. Some cells (macrophages, polymorphonuclear cells) appear to pass from the sinuses to the medullary cords. The size and cell composition of these cords vary profoundly during antigen stimulation.

Billroth cords resemble medullary cords except that they are not organized around vessels but along capillary sinusoids in the red pulp of the spleen, lying thus adjacent to macrophages degrading blood cells.

Thymus-Dependent Areas. T-Dependent areas are lymphoid zones present in all peripheral lymphoid organs: paracortex in the lymph nodes,

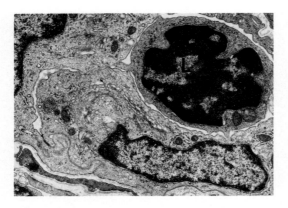

Fig. 3. Lymphocyte (L) passing between 2 endothelial cells of an HEV. ×6,300.

periarteriolar sheaths in the spleen, interfollicular zones in the tonsils, appendix and Peyer's patches. They regress after thymectomy or do not develop in nude mice [23]. These areas are populated essentially by T cells (lymphocytes and immunoblasts), but contain also macrophages, dendritic cells, fibrocytes and migrating B cells [24]. Among these, the dendritic cells appear to play a central role: called interdigitating, lymphodendritic or Steinman's cells, they derive from monocytes; some of them arise from epithelial Langerhans cells and from veiled cells [25, 26]. Dendritic cells which are strongly MHC-class-II-positive fix antigen, and process and present it to T cells.

Other essential components of this T-dependent area are the high endothelial venules (HEVs; fig. 3), which are not present in the spleen. HEVs express surface molecules recognized by leukocytes, namely lymphoid cells [3]. The T and B cells expressing the correct receptors can fix on the endothelial cells, traverse the endothelium across the basal membrane and gain lymphoid tissues where they preferentially move in subnodal spaces, as reported by Sainte-Marie and Peng [27]. This phenomenon is essential for the recirculation of the lymphoid cells. Virgin B cells apparently express receptors for all HEVs, whereas memory cells do not express them and are restricted to defined recirculation zones [3].

Marginal Zone. Inside the spleen, the transition space between the white and red pulp forms the marginal zone. This ill-delimited zone contains fewer

lymphoid cells than the white pulp but more than the red pulp. It surrounds the follicles and the periarteriolar sheaths. Cell types composing it
are T and B cells, macrophages and dendritic cells. Many of these are
migrating cells, either towards or from the white pulp; others are resident,
especially B cells (IgM$^+$, IgD$^-$) which seem to react to T-independent antigens [28–30]. Coronal B lymphocytes and marginal B cells constitute two
separate lineages [31].

The physiological phenomena occurring in this zone are of importance
since the central arterioles, when leaving the white pulp, divide there in
penicillate arterioles which open in the sinusoids or directly in the reticular
meshwork. The marginal zone seems thus to be a starting point for immune
reactions. Depending on the degree of stimulation of the immune system, the
size of the marginal zone varies notably. According to Claasen et al. [32],
marginal-zone macrophages are essential for those B cells responding to
T-independent antigens; Kraal et al. [30], however, demonstrated that these
macrophages are not involved in this type of response but mainly in the
phagocytosis of blood-borne bacteria.

Structures strictly corresponding to the marginal zone have not been
found in other lymphoid organs; however, in the lymph nodes, the interfollicular areas beneath the subcapsular sinus exhibit similar features (cell
composition and arrival of antigens). In the MALT, the subepithelial zone
(above the lymph follicles) may also be its equivalent.

Lymphocytic Infiltrations. Infiltrations of lymphocytes may occur beneath epithelia and around blood vessels, especially in damaged tissues.
Such accumulations of lymphoid cells are diffuse and without organization
in lymph follicles or T-dependent areas. These infiltrations are, however,
not stochastic since the cell types encountered in a given area present
peculiar phenotypes which are not found elsewhere. For example, in the
MALT, suppressor T cells are more numerous in the intestinal villi than
helper T cells; the latter are more abundant in follicle-associated epithelia
[33].

Inflammatory accumulations develop in response to local stimuli (antigens, toxins); cytokines or other mediators then induce vasodilation, cell
migration, nidation and activation. Polynuclear cells, mast cells and monocytes share in this phenomenon.

Lymphoepithelial tissues like those found in tonsils are generally classified as part of the 'lymphoid-cell-infiltrated tissues'. Cell types composing
these zones are epithelial cells (squamous epithelium in the crypts of the

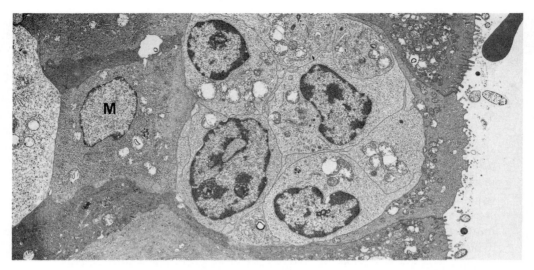

Fig. 4. M cell in the subfollicular dome of a mouse Peyer's patch. This cell harbors various lymphoid cells and favors the transfer of particulate antigen from the intestinal lumen to the immune system. ×4,300.

palatine tonsils), M cells (fig. 4), T cells among which α- and β-receptor-positive cells are found, B cells, plasma cells, macrophages and dendritic cells. The lymphoepithelial tissue of the tonsillar crypts is well organized and cannot be regarded as occasional infiltrations of the epithelium, but as a structured tissue closely related to the development of the subjacent tonsillar organization.

Influence of the Nervous System

All lymphoid tissues are in close relationship to the nerve endings which appear to act principally on the local blood vessels but apparently act also directly on the lymphoid cells via neuromodulators. On the one hand, human lymphoid cells express receptors for β-endorphin, somatostatin, substance P and vasoactive intestinal peptide [34, 35]; on the other hand, the hormones oxytocin and vasopressin, considered normally to be produced by the nervous system, are secreted in the lymphoid organs [36]. It seems that reciprocal actions between the lymphoid and nervous systems occur, but additional studies must clarify their functional importance and mechanisms.

Microenvironments for B Cell Activation

Antigen Penetration Sites

With the exception of the autologous antigens produced by the body itself through mutations or abnormal metabolic processes and by retroviruses, the antigens invade the organism via routes where the natural immunity is weakened or overwhelmed by the number of microbial agents or parasites. The skin, the digestive tube, the respiratory ways and the genitourinary tract are penetration sites of antigens. Wounds, local irritations and epithelial desquamation, produced accidentally or provoked by pathogenic agents, are openings for exogenous products. Once entered, the antigens are concentrated at sites rich in macrophages, e.g. liver, lung, lymph nodes and spleen. This allows the histiocytic defense system to destroy the major part of the foreign material. Some specialized cells focalize, process and present a fraction of this material to lymphoid cells (Langerhans cells and related dendritic cells [37, 38]).

In the gut, the M cells of the follicle-associated epithelia (fig. 4) are specialized to transfer antigens to the macrophages or lymphoid cells lodged in their cytoplasmic folds [39, 40]. These cells transfer particulate or undegraded antigen, whereas the other epithelial cells, which are MHC-class-II-positive after cytokine induction, can present processed material to lymphoid cells [33, 41].

It has been demonstrated that B cells, even quiescent B cells, present antigen to T cells [10, 42, 43]; however, according to Inaba and Steinman [44], dendritic cells are always required. The stimulation of helper or suppressor/cytotoxic T lymphocytes can occur at the antigen penetration site, for example, in the skin where Langerhans cells, special dendritic cells or keratinocytes interact in yet obscure pathways [45].

The route by which an antigen invades the organism is of importance since intravenous, intramuscular and subcutaneous antigen injections will mainly lead to IgG production whereas enteral administration will induce tolerance by IgA production [46]. Since allergic, hypersensitivity or immunologically induced environmental diseases have been the focus of many studies in recent years [47–49], it appears necessary to further improve our knowledge of the antigen or allergen penetration sites and of the environmental factors altering the barriers against foreign substances.

Activation Sites of B Cells

During the first days of a primary antigen stimulation, bone-marrow-derived virgin B cells are activated; thereafter, the recruitment of responsive cells is made in the memory B cell pool [7, 50–52]. Activation consists in cell enlargement, and expression of MHC antigens and other surface molecules [53–56; see also chapter 1 of the present volume]. More particularly, the expression of low-affinity IgE receptors (CD23), depending on the IL-4 production, characterizes the B cell activation [57, 58]. The activation of B cells occurs at those places where antigens as well as additional stimuli arising from other cells are available. The control of the B cell activation by T or accessory cells is complex and mediated by bioactive factors [59, 60].

Even if the cells can contact an antigen at the penetration sites, the activation must occur at those locations where also T cells are stimulated. The exact places of this process are not known. The most opportune sites are the T-dependent areas, since cells loaded with antigens (macrophages, especially dendritic cells, veiled cells, Langerhans or interdigitating cells) concentrate there and since T and B cells invade it through the HEVs (fig. 1). Interestingly, Kraal and Twisk [61] found a specific B cell retention in draining lymph nodes.

The scenario could be as follows: Langerhans cells, having picked up and processed antigen in the epithelium, gain the lymph (veiled cells) and settle in the paracortex of the draining lymph nodes (interdigitating cells [62]). There, this arrival of highly MHC-class-II-positive cells would perturb the local equilibrium, activate T cells and produce bioactive molecules (e.g. IL-1, tumor necrosis factor, lipoxygenase and oxygenase products) which will modify the metabolism of the other cells and induce counteractions. Inaba and Steinman [44] described cell clusters which could correspond to these stimulation sites. In consequence of the cytokine production, the physiology of the vascular cells will also be changed: generally a higher blood flux follows, but also the expression of receptors in the HEVs is increased, inducing a more pronounced transit of cells [63]. Recently, Steinman and Inaba [64] proposed that dendritic cells move and literally find the right T cell clone. We suggest that, since T and antigen-presenting cells have migratory capacity, they will encounter each other in those histological tissues where they settle preferentially. The T-dependent areas exhibit all histological features favoring this event and, thus, are pivots where cells encounter each other and antigens. They are adapted to face the perturbations caused by the antigen penetration (vascularization, reticular connective tissue, lymph flux, cell traffic).

We can also state that the primary lymph follicles are not the activation sites for virgin B cells; first, antigen does not penetrate into the primary lymph follicles at the beginning of the stimulation [27] and, second, primary follicles do not contain activated T cells. Even secondary follicles do not appear to be virgin B cell activation sites since, inside the GCs, these IgM- and IgD-positive cells are rare. Moreover, virgin cells bear Fc receptors (in contrast to memory cells [5]); they would, thus, be inhibited by the immune complexes [65] present on FDCs. According to Gray [66], the recruitment of virgin B cells occurs also outside the follicles. Kroese et al. [67] reported that GCs are oligoclonal, and the absence of virgin B cell activation therein may partly explain this restriction.

Many controversies exist as concerns the requirements of direct contacts between B and T cells. It has been established in culture systems that B cells can be activated by T-cell-derived factors. In vivo, it seems that virgin B cells need a direct contact with T cells, whereas memory B cell activation occurs under lesser stringent T cell control [68]. The high positivity of class II antigens on GC cells is puzzling, especially if these cells do not require direct contact for activation with T cells. Moreover, antigen is present in GCs in the form of immune complexes fixed on FDCs. Perhaps, its presentation to T cells inside the follicles is not followed by a T cell proliferation, being thus different from the process occurring in the T-dependent areas. It is also worth noting that not only MHC class I or II molecules are required during the contacts between the cells of the defense system but that also other molecules as CD4 and integrins [64] play a role. This difference in the requirement of T cell help between virgin memory B cells leads to the hypothesis that the activation of memory cells can occur at more and different locations than that of virgin B cells. This could be the reason why established immune responses are maintained by clonal reactivation soon after immunization and not by continued incorporation of new clones [66].

The mode of presentation of antigen to B cells by macrophages is still unclear. We know that B cells recognize antigen in its native state [37]; however, Rizvi et al. [69] demonstrated that dengue virus must be processed by macrophages (class II⁻ or class II⁺) for the activation of B cells. The activation of T-independent B cells may occur at the antigen penetration sites, but, apparently, the main area for their stimulation, proliferation and differentiation is the marginal zone of the spleen where these IgM-positive, IgD-negative and T-independent reactive B cells are concentrated [28].

Microenvironments for B Cell Proliferation

Lymph Follicles
At the beginning of the primary response, GC precursor cells (virgin cells) appear to be IgM- and IgD-positive cells with low-affinity immunoglob-

ulins but a broad range of antigen recognition capacity. Once activated, they lose IgD and their receptors for HEVs [62, 70] before or during their penetration into the lymph follicles. Inside the GCs, they express glycoproteins with affinity for peanut agglutinin (PNA) [62] and particular adherence molecules (integrins).

During the first days of stimulation, B cells are proliferating blast cells (centroblasts concentrated in the dark zone; fig. 1). When some of them stop dividing they accumulate in the light zone (centrocytes). The proliferation rate is very high; apparently the cell cycle lasts for about 6 h [19]. The cell degeneration is, however, intense in regard to the activity of tingible body macrophages which phagocytose numerous cells, even dividing cells [71, 72]. The tingible body macrophages appear early during the GC reaction [25]; the reason of this high elimination rate of lymphoid cells has been discussed elsewhere [16].

The GCs are apparently oligoclonal, which could be accounted for by several reasons: restriction of immigration in the GC; limitation of stimulation to certain clones (resulting either from a helper T cell selective action or from a specific antigen reaction); selective phagocytosis by tingible body macrophages, and/or prohibition of virgin B cell stimulation by immune complexes. The microenvironment for B cell proliferation in the GCs is unique since such particular cells as FDCs, GC T cells and tingible body macrophages are found nowhere else. The main features of these cells are summarized in the following sections.

Follicular Dendritic Cells. Inside the follicles and more particularly in the GCs, the FDCs (which should not be confused with the dendritic cells of the T-dependent areas) form a peculiar stroma. FDCs develop apparently from reticular fibroblasts [73–75]; nevertheless, since FDCs exhibit some surface antigen in common with myeloid cells, Schriever et al. [76] consider them as having a unique phenotype. Some authors [77] suggested that several FDC types exist. During ontogeny, the formation of primary follicles is only possible when FDCs are developed [78].

FDCs extend long and complex cytoplasmic projections through the GC and even in the inner part of the corona. These extensions bear numerous dendrites and lamellae [79, 80]. Adjacent projections interpenetrate and are linked by desmosomes [81, 82]. This cytoplasmic meshwork envelops the lymphoid cells (fig. 5) and extends principally in the light zone; in the dark zone, the FDCs appear to be pushed aside by the dividing centroblasts. On the surface of these cytoplasmic extensions, immune complexes are fixed

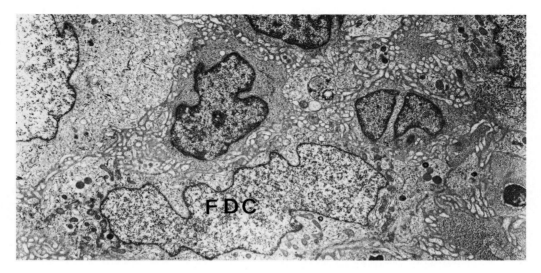

Fig. 5. FDC in the light zone of a GC. It surrounds lymphoid cells with its cytoplasmic projections. ×8,000.

[83, 84] by C3b receptors [85–87] or by Fc receptors [88, 89]. Free antigens are not retained by FDCs; consequently, during the primary stimulation, antigen only localizes in the GCs when antibodies are produced [90]. Phagosomes or lysosomes are rarely seen in FDCs [81, 82], and no clear endocytosis of immune complexes could be observed. These complexes can, however, be retained for long periods (several weeks to months [91]). All immunoglobulin classes have been identified on FDCs, except for IgD [86, 89].

Subcutaneously or intravenously injected immune complexes are retained after few minutes in the draining lymph nodes or in the splenic GCs, respectively [81, 88]; these complexes displace those fixed previously [92, 93]. Many virus particles lodged inside the FDC dendrites are retained in form of immune complexes or are produced by FDCs. These observations are of importance since HIV particles have been repeatedly observed in or on FDCs [94, 95].

Different modes of entry in lymph follicles have been proposed for antigen, antibodies or immune complexes: simple diffusion [96]; local production of antibodies; transfer along cellular projections [97], or transport by cells [98–101]. Our observations indicate that, under normal physiological

Table 1. Antigens localized on FDCs

Category	Antigens
FDC-specific antigens	human: R4 123 (DRC1) [199]
	rat: ED5 [77]
Lineage-restricted antigens	myeloid lineage: CD11b, CD14, CD16, CD31
	lymphoid lineage: CD19, CD21, CD23, CD24, CD37
Nonlineage-restricted antigens	CD37, CDw40, CD71
	MHC class I and II
Adhesion molecules	VLA-3, ICAM-1, CD11b
Exogenous material	IgA, IgE, IgG, IgM; complement factors;
	S-100 protein; different antigens (e.g. virus particles)

The data are taken from references 76, 86, 87, 89, 90, 104, 105, 110, 195, and 199.

conditions, the immune complexes are rarely free in the lymph or intercellular liquids; they are cell-bound or endocytosed. B lymphocytes appear to be the best candidates for the transfer of these complexes to FDCs, since they can fix them, attach to FDCs and then hand them over to FDCs [102, 103]. Fc receptors appear to play a determinant role during this transfer, especially since lesser immune complex trapping occurs on FDCs after lipopolysaccharide treatment which alters the expression of Fc receptors on lymphoid cells and on FDCs [102].

Multiple antigens have been identified on FDCs (table 1) [104]. Johnson et al. [105] divided them into three categories: intrinsic antigens found only on FDCs; intrinsic antigens shared with other cells or tissue constituents, and exogenous antigens acquired passively by FDCs through surface receptors.

Enzymatic techniques demonstrated solely a high membrane 5-nucleotidase activity on FDCs [106].

After antigen stimulation, the cytoplasmic FDC extensions increase in size and number [90]; the proportion of FDCs to other GC cells is about 2% [107] and they occupy near to 13% of the GC volume [108]. When isolated by mild enzymatic dissociation methods [107, 109] FDCs remain in close contact with lymphoid cells originating from the GCs [110, 111]. Recently, Schriever et al. [76] obtained pure FDCs by using a cell sorter. The contacts FDCs establish with lymphoid cells are probably mediated via adhesion molecules (ICAM-1, VLA), immune complexes and complement factors, and depend on the presence of bivalent cations and the temperature [76]. Thus,

FDCs create a particular microenvironment for lymphoid cells and are found nowhere else. They influence the lymphoid cells in improving their survival and proliferation [112, 113], but apparently also in reducing their differentiation into antibody-secreting cells (see below). According to Lortan et al. [114], the follicles are the most obvious sites for continued antigen supply for B cell reactivation. The 5-nucleotidase of FDCs can play a role at different levels: it may discard nucleotide monophosphate molecules which are toxic [115]; deliver nucleotide molecules to lymphoid cells which possess receptors (e.g. for adenosine [116]), or set out nucleotides which can induce vasodilation [117].

Even if FDCs do not play a role during the virgin B cell activation (no antigen), they sustain the B cells during their divisions by their direct contact with lymphocytes or in releasing cytokines [112, 113], by presentation of C3b (B cells have CR1 receptors [118]) or, on the contrary, in removing Fc receptors from B cells (Fc receptors are inhibitory [65]).

Germinal Center T Cells. T cells represent about 5% of the GC cells [111, 119]; most of them are CD4-positive [120, 121]. According to Kroese [122], GC T cells exhibit simultaneously the phenotype of helper and suppressor T cells in rats and have the Leu-7 antigen in common with natural killer cells. Recently, Katz et al. [123] emphasized that natural killer cells can enhance the proliferative response of B lymphocytes. These T cells are sedentary (Mel 14-negative [70]), bear α, β-receptors, and apparently do not produce IL-2 [124]. Activated T cells are rare in GCs [125]. Even after isolation, GC T cells preserve their contact with B cells, macrophages and FDCs. They appear like small, nonactivated lymphocytes. Their role is essential for the GC reaction but the way they act is unknown [126]. T cells bear Fc receptors for different immunoglobulin classes and appear to regulate B cells (proliferation, isotype switch, immunoglobulin production [127]); maybe that FDCs, in retaining immune complexes, present these complexes (not via MHC antigens) to T cells and influence in this way the T cell control of B lymphocytes. When their function is perturbed, for example during HIV infection, severe alterations of the GC morphology appear [128].

Tingible Body Macrophages. Tingible body macrophages appear soon after the GC reaction; they phagocytose high numbers of lymphoid cells, including those in the S phase [24, 71]. We have discussed previously the reason for this endocytosis of lymphoid cells [16]. Tingible body macrophages express no or only low levels of MHC class II molecules, Fc or C3b receptors; in consequence, they seem not to exert a noticeable regulatory effect on T cells. However, we consider that, in degrading cells, they set out large quantities of molecules useful for the dividing B cells. Thus, they form a microenvironment where they eliminate degenerating or abnormal cells but favor the survival and proliferation of other cells.

Besides these macrophages which do not absorb antigens, GCs also contain normal macrophages loaded with antigen [129]. Endocytosis of gold-labeled immune complexes by these macrophages was observed in the draining lymph nodes several days after an injection [unpubl. data]. The role of these normal macrophages is unclear: they seem, however, not to present antigen to the GC T cells since these latter appear like small lymphocytes [125] and do not produce IL-2 [124].

Vascularization, Connective Tissue and Bioactive Molecules. Numerous capillaries are observed in secondary follicles. They are surrounded by pericytes and FDCs positive for alkaline phosphatase [130]. Fibroblasts and collagen bundles are found mainly between the corona and the GC. Dark cells appear as heterogeneous degenerating cells [131]. Type III collagen is the main component of the reticular fibers and is associated with other elements (proteoglycans, aminoglycans, glycoproteins) which remain to be analyzed. It has now been evidenced that all these components play a role in the behavior of the cells (adhesion, cell movements, homing), especially since it was observed that many surface molecules (very late antigens, integrins and others) can fix on them [132, 133].

Data on cytokine production by GC cells are rare. Velardi et al. [124] showed that GC T cells do not produce IL-2 in culture. We have reported that GC cells obtained from human tonsils and cultured in the presence of lectins produced IL-1β, tumor necrosis factor-α and IL-6 [113]. Neither IL-2, interferon-γ nor IL-4 were detected, but they could have been consumed by the cultured cells. We could not demonstrate a direct secretion by FDCs. These latter may improve the secretion of cytokines by other cells; indeed, the level of the secreted cytokines was higher when FDCs were present [113]. Using the same culture system with isolated GC cells, a secretion of prostaglandin E_2 [134] but not of histamine [unpubl. data] was observed.

Using in situ hybridization technique, we could localize the IL-6-producing cells in the human tonsils; they are dispersed in the subepithelial and T-dependent zones, but not in the GCs [135] (fig. 1). Schriever et al. [76] also concluded that FDCs do not produce IL-6. The GCs apparently harbor

IL-5-secreting and -sensitive cells [136]. Studies on these cytokines must be refined since most of them are pleiotropic in their effects, which is due to their interactions and to the state of responsiveness of the target cells.

Nonfollicular B Cell Proliferation Areas

Lymph follicles are not the sole sites where antigen-stimulated B cells divide. T-dependent B cells can apparently proliferate in the MALT outside the follicles, but clear data are lacking. We have already mentioned the T-independent B cells which proliferate in the splenic marginal zone [28, 29]. Although being actively studied, these CD5-positive cells (Ly-1$^+$ in mice) remain a puzzling cell population, belonging to the natural immunity system and behaving in a different manner than the classical B cells which ensure the acquired immunity.

CD5-positive B cells appear early during phylogenesis, derive from the bone marrow stem cells and mainly defend the organism against microbes. Their surface immunoglobulins result from gene rearrangements, but some variable segments of the light chains are repeatedly expressed in clonally unrelated cells. These immunoglobulins are multispecific even after antigen stimulation (no hypermutagenicity). CD5-positive B cells are responsive to dextran sulfate, are the major source for chronic B cell leukemia and are frequently autoreactive [68, 137–139]. These B cells occur in the peritoneal cavity, the blood and the spleen where they proliferate in the periphery of the periarteriolar sheaths and differentiate in the vicinity of metallophilic macrophages [140]. Some IgM- and IgD-positive cells populate the deep areas beneath the subcapsular sinus in the lymph nodes, where, however, poor T-independent responses have been detected [141]. Microenvironments specific for these cells have not been examined.

Microenvironments for Memory B Cell Formation

Memory cells can be harvested a few days after the primary stimulation. Their existence has been shown by transfer into irradiated animals. Memory B cells exhibit several characteristic features (see introductory section) and are formed in sequential events, probably in relation to appropriate climates.

Clonal Amplification

The GC is the site of antigen-driven clonal amplification. At the beginning of the primary response, virgin cells (generally multispecific and with low affinity) are recruited and induced to divide. During the course of this primary response, high-affinity clones dominate by their capacity to react and proliferate [66].

Hypermutation

The hypermutation phenomenon, which changes the affinity of the immunoglobulin paratope at the level of the variable heavy-chain regions, has been well described [68]. The parameters conditioning it are not known. It develops after an antigen stimulation and apparently occurs in the GCs. The somatic mutations are introduced stepwise during the clonal proliferations [142]. The exact site of hypermutagenicity remains to be determined; it may happen in the dark zone where cells are dividing. In consequence of these somatic mutations, some cells lose their reactivity to antigen, others remain unchanged (silent mutations) and some present a higher affinity. These latter will probably be favored, proliferate and differentiate further. Poorly reactive or nonreactive cells apparently disappear (maybe subsequently to a negative action of the stimulated cells) and are phagocytosed by the tingible body macrophages. The mutated cells are able to expand and to terminally differentiate upon a novel antigen encounter. It is not excluded that some of them may again enter the pathway of somatic hypermutation [143].

Isotype Switch

Whereas some immunoglobulin class switch may happen in the chorion of the MALT system, strong evidence has been given that this event occurs inside the GCs [3, 144]. Indeed, a change in the immunoglobulin heavy-chain isotype expression is seen during the primary antigen stimulation. IgM is expressed first, then IgG in the lymph nodes and spleen, and IgA in the MALT. Chronologically it seems to follow the somatic mutations described above and turn them off irreversibly [145–148]. T cells or factors they secrete determine the immunoglobulin switch. Especially IL-4 and IL-5 act on cells which have performed the immunoglobulin class switch; this has been extensively studied in the case of IgE [149, 150–153].

Coulie and Van Snick [154] showed that injections of immune complexes composed of defined immunoglobulin isotypes can induce the expression of other isotypes. Since FDCs retain immune complexes and since the switch occurs in the GCs, it is tempting to speculate that FDCs, in presenting these complexes to B or T cells, influence the immunoglobulin switch. In this context it is worth mentioning that the IgE isotype expression on Peyer's patch cells can be modulated by feeding [151].

Loss of Fc Receptor Expression

Virgin B cells express several Fc receptors on their surface; memory B cells and plasma cells do it lesser or not [155]. Here too, microenvironmental conditions may play a role. We have shown that lymphoid cells can transfer immune complexes to FDCs [102]. It is not excluded that this exchange also concerns the whole or a part of the receptor itself.

Recirculation Molecules

Once activated, B cells loose their vascular addressins, but when differentiating into memory cells they reexpress them in a more restricted fashion. Memory B cells recirculate in defined areas inversely to virgin cells which bear the addressins for all HEVs [156–159]. The determinants for this restriction are unclear; cells (epithelial or others) of the zone where the memory cells differentiate and reacquire addressins may play a role in this mechanism.

Transformation in Small Lymphocytes

The events occurring inside the GC are obscure. It seems that centroblasts, when passing to the light zone, transform into centrocytes and then into small lymphocytes which apparently leave the follicles through the corona to gain the efferent lymph or blood vessels (fig. 1). However, according to several authors [19], centrocytes can also give centroblasts. All these cells are PNA-positive. The circulating memory B cells become low PNA-positive (lymphocytes), but can reexpress it at a high level after a novel antigen stimulation [160].

The elements of the light zone of the GC act in a yet ill-defined manner in this evolution. Since the cytoplasmic FDC extensions, which are covered with immune complexes, extend mostly in this area [90, 161], they appear to build up an appropriate microenvironment for the memory B cell formation. We observed lesser immunoglobulin production in lymphoid-cell cultures when FDCs were present [112]. It may be that FDCs inhibit the transformation into immunoglobulin-secreting cells, thus favoring the evolution of centrocytes to small memory lymphocytes. Immune complexes inhibit the terminal differentiation of B cells [162, 163]. However, it may also be that FDCs push the cells toward the memory maturation and thus, indirectly, reduce the chance of B cells to give plasma cells (see also below). The balance between memory cell and plasma cell formation depends on the local conditions (e.g. antigen, antibodies, cytokines), thus on the microenvironment at a given time. Since after irradiation, which induces depletion of lymphoid cells, many plasma cells develop even in GCs, it can be speculated that lymphoid cells prevent the terminal differentiation in normal conditions, perhaps under the influence by cells like FDCs.

According to Gray [50], memory B cell populations have a half-life of a few weeks in the absence of antigen. When antigen is available, B cell clones undergo repeated stimulation and may be very long-lived. Resting memory B cells themselves are thus not long-living cells. Besides of being able to proliferate, memory B cells can transform into plasma cells either shortly after their exit of the follicles or after a long period of recirculation. This transformation depends on the local conditions (e.g. IL-6, IL-2) they encounter during their migration. For example, in the tonsils, cells leaving the follicles can meet IL-6-producing cells in the subepithelial zone or in the interfollicular space [135] and be induced to transform into plasma cells (beneath the epithelium) or gain the efferent lymph and recirculate before colonizing a mucosa (fig. 1). A proportion of the memory B cells produced in the spleen following a T-dependent antigen stimulation can settle in the marginal zone, others enter the recirculating memory B cell pool [164]. The mechanism of this process is obscure.

The fate of the small lymphocytes is thus not simple. They exhibit morphological and antigenic differences and react according to their constantly changing environment. During their circulation, lymphocytes spend a lot of time in the blood and lymph. These media do not allow a proliferation or differentiation, neither are innocuous since they carry endogenous molecules (e.g. hormones, cytokines, immunoglobulins, complement factors, immunoglobulin-binding factors) or exogeneous substances (e.g. bacterial products, nutritional elements, drugs). All these agents vary in concentration with time or locally and affect the behavior of the cells.

Microenvironments for Terminal B Cell Differentiation

In lymph nodes, the final maturation from a centroblast of the dark zone toward a plasma cell in the medullary cord lasts for about 24 h. During this period of time, the cell leaves the follicle, traverses the paracortical zone, migrates along blood vessels and settles in the vicinity of the medullary sinuses [165]. Thus, it moves through different microenvironments. Plasma cells do not, or only at a low level, express MHC class II molecules, Fc and C3b receptors, B cell antigens and surface immunoglobulins; a pressure thus exists on the gene regulation of the plasma cells. Many genes are repressed, while others are expressed as for example the PC1 and PC2 surface molecules [166]. Plasma cells have a restricted life span (2–8 days [19]) but some of them can be reactivated [165]. The immunoglobulins are apparently secreted without storage, but pathological accumulations of intracytoplasmic immunoglobulins have been described [for review, see ref. 19].

Most B cells do not transform into plasma cells in draining lymphoid organs; they home to the bone marrow, mucosal tissues or other defined sites, or continue to recirculate in form of memory cells. The local climate of an organ regulates the transformation into antibody-secreting cells and influences the lifespan of the plasma cells. Since different tissues are likely to harbor plasma cells, we describe them now more in detail.

Lymph Follicles

Primary lymph follicles do not contain plasma cells. Even in the GCs, no or only a few number of plasma cells are present [167, 168]. This environment appears thus not to be adequate for a terminal differentiation of B cells. Perhaps, FDCs repress this transformation or do not produce the adequate factors [112]. Among the cytokines IL-6 is reputed for induction of immunoglobulin secretion. Our laboratory, using in situ hybridization techniques, could not detect IL-6-producing cells in lymph follicles [135]. These results confirm those obtained by Schriever et al. [76] with isolated FDCs. Nevertheless, several authors observed the presence of antibody-secreting cells in the GCs [19, 169, 170]. It appears to be of minor importance [171]. The most apparent function of the GCs is the production of memory cells. An internal equilibrium seems to exist there between pro and contra forces pushing B cells to divide and to differentiate into memory cells and preventing them from an immediate transformation into antibody-secreting cells. When this equilibrium is disrupted, plasma cells can apparently develop. The analysis of Castleman's disease, which is characterized by an excess of antibody production and lymph node hyperplasia, supports the hypothesis of a GC disequilibrium. Indeed, in some cases of Castleman's disease, Yoshizaki et al. [172] showed an excessive IL-6 production which perturbs the physiological status of the GCs and thus induces a hyperproduction of plasma cells.

B cells can leave the GC following two routes. One passes from the centroblasts and centrocytes toward the lymphocytes which traverse the corona to gain the lymphoid sinuses or blood vessels (fig. 1). The second passes from the centroblasts to the base of the follicles toward the zones where the cells differentiate into plasma cells. Which pressure inside the GC pushes the centroblasts in one or the other direction is unclear; maybe the vicinity of FDCs bearing immune complexes or that of the T-dependent zone influences this evolution.

T-Dependent and Annexed Areas

Plasma cells are not found in the splenic periarteriolar sheaths nor in the paracortical zones of lymph nodes. B cell cohorts traverse them or migrate along them. The cells of these zones exert possibly a positive action on B cells in pushing them to their terminal differentiation which, however, is completed in adjacent areas (medullary cords, Billroth cords, subepithelial zones). It is well known that T cell factors and monocyte/macrophage products are necessary for this transformation (e.g. IL-2, IL-4, IL-6, tumor necrosis factor [173]).

The mucosae along the gut, the respiratory system and the genitourinary tracts appear to be privileged places for plasma cell formation. It is worth noting that the gingiva [174], and the salivary, lacrimal and mammary glands are particularly rich in plasma cells. The intestine is the main producer of IgA [33]. The reasons for that appear multiple; HEVs seem to express peculiar surface molecules responsible for the specific homing of cells [156]. However, the local release of antigens and the production of epithelial factors (secretory component and others) can also play a role in this oriented migration of cells [175, 176]. In absence of antigens (axenic mice) fewer inflammatory and plasma cells populate the intestinal mucosa [177]. Plasma cells aggregate around the crypts or inside the villi. CD4-positive cells are frequent in both these sites [178], and seem to control the immunoglobulin secretion and the isotype switch [179, 180]. Staite et Panayi [181] reported a stimulation of the immunoglobulin production by prostaglandin E_2. This observation may be of importance since several cells of the lamina propria produce prostaglandins.

Groups of plasma cells are also found adjacent to the marginal zone in the spleen [169]. This aggregation is probably dependent on local stimuli which may be similar to those determining the plasma cell formation in the bone marrow (see below).

In the lymph nodes, plasmoblasts of the medullary cords migrate along the blood vessels [18]; these cells contribute to the maturation most probably in supplying with nutritional elements. However, it is also possible that signals arising from the serum are transmitted to these immunoglobulin-secreting cells and influence the rate or the type of immunoglobulin produced. Furthermore, medullary cords harbor many polynuclear cells and mast cells (fig. 2); the effect of these cells on the antibody production is unclear but is worth being studied.

Bone Marrow

The bone marrow is the main site of immunoglobulin production after the digestive tract. According to the reports of Hijmans et al. [182] and Mac Millan et al. [183], more than 90% of serum immunoglobulin is produced by the plasma cells of the bone marrow. No antigen-stimulated B cells proliferate there; thus, the bone marrow retains a high number of circulating memory B cells which transform into plasma cells [19, 184, 185].

Different classes appear to be produced; in mice, IgGl, IgG2, IgG3 and IgA are the major classes secreted. The number of producing cells in the bone marrow increases with aging (35,000 in 4-week-old mice to 5×10^5 at 14 weeks and 3×10^6 at 2 years [184]).

The microenvironments inside the bone marrow are too ill-defined to give accurate clues for this immunoglobulin production. The reticular tissue, the macrophages and blood cells, and the production of bioactive molecules would explain the transformation of B cells into plasma cells, but the fine-tuning mechanisms of this development are unknown.

We tried to find common features explaining the locations of plasma cells in the mucosae, bone marrow, spleen and lymph nodes. The histological tissues allowing the terminal B cell differentiation appear to be quite variable: e.g. vicinity of epithelia (intestine, tonsillar crypts); location around blood vessels (splenic sinusoids, bone marrow), and nearness to lymph sinuses (lymph nodes). A lot of tissues do not harbor plasma cells (nervous tissue, cartilage, compact bone, eye, ear), and others are not specialized in immunoglobulin production (liver, endocrine glands, testicles, ovary). B cells are thus not allowed to transform at any given place into plasma cells. We suggest that a repression mechanism exists which prevents B cells from the terminal differentiation, but that B cells, when migrating through particular zones (T-dependent), may receive signals for terminal differentiation which can be achieved at distance in locations where additional signals are given, possibly by epithelial or other cells (mast cells, polynuclear cells, macrophages, natural killer cells), affecting settling, excretion and life span.

Influence of B Cells on Their Microenvironment

B cells are not simply influenced by their surrounding elements, they effect also profound modifications in their environment. During stimulation, the local accumulation and clonal growth of B cells involve tissue expansions easily detectable at the level of the spleen and the lymph nodes. They metabolize and reject products (for example ATP derivatives), consume, e.g. nutrients, cytokines, complement factors and hormones, and even set out suppressive factors [186]. The surface molecules they express (MHC molecules, integrins, immunoglobulins) influence their neighboring cells. The immunoglobulins secreted exert multiple actions on different systems, as for example on blood components, either on the complement cascade or on the cells expressing Fc or complement receptors (polynuclear cells, red blood cells, platelets, monocytes), and on endothelial and epithelial cells. Especially the function of T cells appears to be modulated by the different immunoglobulin isotypes.

A lot of contradictory data have been reported as concerns the immune complexes [187–189]. They exert suppressive effects on the immunoglobulin production [162, 190] and on Ia expression [191], and influence the immunogenicity of the antigen [192]. The mode of action of the immune complexes depends on the ratio between antigens and antibodies, the immunoglobulin isotype and affinity, their concentration, their location, the activation of the complement cascade, and so forth. The situation is much too complex to be presented here.

Altered Microenvironments and Pathologies

Although the relationships between abnormal changes of the microenvironment and the generation of pathologies have been described in specialized reviews, we will briefly summarize the impacts of the alteration of tissue or cellular processes on the development of abnormal or deficient immune reactions. Coherent immune responses require fine-tuned cellular interactions in defined locations. Even slight defects in cellular physiology or products can lead to modifications extending from transient atopic reactions to long-lasting hypo- or hyperreactions.

Analyses of genetic defects revealed that a high and varied number of factors are involved in equilibrating immune responses. Defective or abnormal expressions of surface molecules, cytokines, immunoglobulins, complement factors and enzymes have been cited. For example, a disorder of leukocyte adhesion deficiency has been recognized in which the synthesis of the β-chain of integrins is abnormal [193]. Deficiencies in the nucleoside metabolism explain inefficient immune reactions [115, 116, 194]. Pathogenicity due to cytokine overexpression has been exemplified by Yoshizaki et al. [172] who showed that Castleman's disease (lymph node hyperplasia with plasma cell infiltration) results from an excessive production of IL-6. Another cytokine tumor necrosis factor appears to be responsible for the vascular pathology during the course of cerebral malaria [49]. Vascular pathologies can also be induced during immune-complex diseases [6, 195].

Acquired deficiencies arise from exogenous elements perturbing the homeostasis of an organism. The most famous is the HIV which alters the T cell physiology. More particularly, the microenvironment of the B cell production in the lymph follicles is altered, apparently in consequence of a defective T cell help and a disorganization of the follicular dendritic network. The FDCs harbor virus particles and apparently produce them [94, 95, 196]. Bacterium-derived elements can stimulate or brake immune responses; lipopolysaccharide, for example, perturbs regulatory pathways in reducing Fc receptor expression [197, 198].

Human activities produce agents whose effects remain to be studied. They can alter the epithelia, fix on cells, induce autoantibodies, modify the cell behavior and give rise to hypo- or hypersensitivity reactions or to uncontrolled proliferation. A large number of pathologies can be ranged in this category. Tumors can also be induced by abnormal signals delivered in modified microenvironments or can develop after the perturbation of the immune system. These aspects have been described in specialized publications [6, 195, 199].

General Conclusions

Antigen fixation on the surface immunoglobulin of a virgin B cell induces profound changes when survening in adequate environments which provide the additional signals required for the subsequent evolution of the target cell. Table 2 summarizes the locations, physiological events and cellular stages which are involved in this antigen encounter. A lot of clues have been determined as regards the B cell response, but many steps remain obscure: the transition pathways of lymphocytes to centroblasts and centrocytes are hypothetical, and the determination of hypermutation or isotype switch as well as the fate of the memory cells must be better studied.

Future studies will allow the progress from the knowledge of an organ or a tissue toward precise data defining the microenvironments of the different cells involved. Fine-tuned cell contacts and communications guide cells in defined directions so that, for example, the 1,000 milliards of lymphoid cells populating a human body act in an equilibrated and efficient manner. The whole adaptive immune system works like a single organ, even though it is dispersed throughout the body and is linked to different lymphoid organs. To achieve this, efficient intercellular communications have been developed and strong mechanisms

Table 2. Schema summarizing the histological, physiological and cytological relationships during B cell production, activation, proliferation and differentiation

Tissue environment	Physiological event	Cells
Bone marrow	lymphopoiesis	lymphoblasts----→
Blood, lymph	recirculation	lymphocytes----→
T-dependent areas	activation of immunopoiesis	→immunoblasts→
GC dark zones	proliferation	→ centroblasts----→
GC light zones	memory-cell maturation	←?—centrocytes----→
Blood, lymph	restricted recirculation	memory lymphocytes→
Special microenvironments	terminal differentiation	plasma cells ←

——→ = Usual pathway; ----→ = unfrequent pathway.

control the cellular reactions. Especially these controls are accomplished in strictly demarcated microenvironmental conditions. These features have to be defined with high precision for each cell type analyzed, and will lead to a better understanding and a more accurate treatment of pathological states.

References

1 Nossal GJV: The two waves of B-lymphocyte differentiation. Ann Immunol 1984; 135:195–199.
2 Hayakawa K, Ishii R, Yamasaki K: Isolation of high-affinity memory B cells: Phycoerythrin as a probe for antigen-binding cells. Proc Natl Acad Sci USA 1987; 84:1379–1383.
3 Vitetta ES, Brooks K, Isakson P, et al: B lymphocyte receptors; in Paul WE (ed): Fundamental Immunology. New York, Raven Press, 1985, pp 221–244.
4 Gallatin WM, Weissman IL, Butcher EC: A cell-surface molecule in organ specific homing of lymphocytes. Nature 1983;304:30.
5 Mathur A, Lynch GR, Köhler G: Expression, distribution and specificity of Fc receptors for IgM on murine B cells. J Immunol 1988;141:1855–1862.
6 Bach JF: Immunologie. Med-Sci. Paris, Flammarion, 1986.
7 Gray D, Mac Lennan ICM, Lane PJL: Virgin B cell recruitment and the lifespan of

memory clones during antibody responses to 2,4-dinitrophenyl-hemocyanin. Eur J Immunol 1986;16:641–648.

8 Cerottini JC, Mac Donald HR: The cellular basis of T-cell memory. Annu Rev Immunol 1989;7:79.

9 Byrne JA, Butler JL, Cooper MD: Differential activation requirements for virgin and memory T cells. J Immunol 1988;141:3249–3257.

10 Sanders ME, Makgoba MW, Shaw S: Human naive and memory T cells. Immunol Today 1988;9:195–199.

11 Hogg N: The leukocyte integrins. Immunol Today 1989;10:111–114.

12 Ham AW, Cormack DH: Histology. Philadelphia, Lippincott, 1979.

13 Reynolds JD, Morris B: The evolution and involution of Peyer's patches in fetal and postnatal sheep. Eur J Immunol 1983;13:627–635.

14 Thorbecke GJ, Asofsky RM, Hochwald GM, et al: Gamma-globulin and antibody formation in vitro: III. Introduction of secondary response at different intervals after the primary: The role of secondary nodules in the preparation of the secondary response. J Exp Med 1962;116:295–310.

15 Groscurth P: Non-lymphatic cells in the lymph node cortex of the mouse: I. Morphology and distribution of the interdigitating cells and the dendritic reticular cells in the mesenteric lymph node of the adult ICR mouse. Pathol Res Pract 1980;169: 212–234.

16 Heinen E, Cormann W, Kinet-Denoël C: The lymph follicle: A hard nut to crack. Immunol Today 1988;9:240–242.

17 Paul WE: The immune system: An introduction; in Paul WE (ed): Fundamental Immunology. New York, Raven Press, 1984, pp 1–22.

18 Veldman JE: Histophysiology and Electron Microscopy of the Immune Response. Groningen, Dijkstra Niemeyer, 1970.

19 Lennert K: Malignant Lymphomas Other than Hodgkin's Disease. Berlin, Springer, 1978.

20 Butcher EC, Weissman IL: Lymphoid tissues and organs; in Paul WE (ed): Fundamental Immunology. New York, Raven Press, 1984, pp 109–127.

21 Butcher EC, Rouse RV, Coffman RL, et al: Surface phenotype of Peyer's patch germinal center cells: Implications for the role of germinal centers in B cell differentiation. J Immunol 1982;129:2698–2707.

22 Matthews JB, Basu MK: Oral tonsils: An immunoperoxidase study. Int Arch Allergy Appl Immunol 1982;69:21–25.

23 De Sousa MAB, Parrott DMV, Pantelouris EM: The lymphoid tissues in mice with congenital aplasia of the thymus. Clin Exp Immunol 1969;4:637–644.

24 Hoefsmit ECM, Kamperdijk EWA, Hendricks HR, et al: Lymph node macrophages. Reticulendothel Syst 1980;1:417–468.

25 Kamperdijk EWA, Raaymakers EM, de Leeuw JHS, et al: Lymph node macrophages and reticulum cells in the immune response: I. The primary response to paratyphoid vaccine. Cell Tissue Res 1978;192:1–23.

26 Thorbecke GJ, Silberberg-Sinakin I, Flotte TJ: Langerhans cells as macrophages in skin and lymphoid organs. J Invest Dermatol 1980;75:32–43.

27 Sainte-Marie G, Peng FS: Thymic cell migration in the subnodular spaces of draining lymph nodes of rats. Cell Immunol 1980;52:211–217.

28 Mac Lennan ICM, Gray D, Kumaretane DS, et al: The lymphocytes of splenic marginal zones: A distinct B cell lineage. Immunol Today 1982;3:305–307.

29 Mac Lennan ICM, Bazin H, Chassous D, et al: Comparative analysis of the development of B cells in marginal zones and follicles. Adv Exp Med Biol 1985; 186:139–144.

30 Kraal G, Ter Hart H, Meelhuizen C, et al: Marginal zone macrophages and their role in the immune response against T-independent type 2 antigens: Modulation of the cells with specific antibody. Eur J Immunol 1989;19:675–680.

31 Nieuwenhuis P: B cell differentiation: B cell subsets and factors, organs or structures involved in B lymphocyte genesis. Adv Exp Med Biol 1985;186:203–211.

32 Claasen E, Kors N, Van Rooijen N: Influence of carriers on the development and localization of anti-2,4,6-trinitrophenyl (TNP) antibody-forming cells in the murine spleen: II. Suppressed antibody response to TNP-Ficoll after elimination of marginal zone cells. Eur J Immunol 1986;16:492–497.

33 Brandtzaeg P, Bjerke K: Human Peyer's patches: Lymphoepithelial relationships and characteristics of immunoglobulin producing cells. Immunol Invest 1989;18: 29–45.

34 Ottaway IH: Evidence for local neuromodulation of T cell migration in vivo. Adv Exp Med Biol 1985;186:637–645.

35 Besedovsky HO, del Rey A: Immune-neuroendocrine network; in Cinader B, Miller RG (eds): Progress in Immunology VI. Orlando, Academic Press, 1986, pp 578–585.

36 Geenen V, Legros JJ, Franchimont P, et al: The thymus as a neuroendocrine organ: Synthesis of vasopressin and oxytocin in human thymic epithelium. Ann NY Acad Sci 1987;496:56–66.

37 Unanue ER, Beller DI, Lu CY, et al: Antigen presentation: Comments on its regulation and mechanism. J Immunol 1984;132:1–11.

38 Hume DA, Doe WF: Role of macrophages in cellular defense; in Heyworth MF, Jones AL (eds): Immunology of the Gastrointestinal Tract and Liver. New York, Raven Press, 1988, pp 23–45.

39 Owen RL, Jones AL: Epithelium cell specialization within human Peyer's patches: An ultrastructural study of intestine lymphoid follicles. Gastroenterology 1974;66: 189–207.

40 Ogawa K, Miyoski M: Intercellular spaces in the lymph nodule associated epithelium of the rabbit Peyer's patch and appendix. Arch Histol Jpn 1985;48:53–67.

41 Bjerke K, Brandtzaeg P: T cells and epithelial expression of HLA class II determinants in relation to putative M cells of follicle-associated epithelium in human Peyer's patches. Adv Exp Med Biol 1989;237:695–698.

42 Rock KL, Benacerraf B, Abbay AK: Antigen presentation by hapten-specific B lymphocytes: I. Role of surface immunoglobulin receptors. J Exp Med 1984;160: 1102–1113.

43 Kakiuchi T, Chesnut RW, Grey HM: B cells as antigen-presenting cells: The requirement for B cell activation. J Immunol 1983;131:109–114.

44 Inaba K, Steinman RM: Antibody responses to T dependent antigen: Contributions of dendritic cells and helper T lymphocytes. Adv Exp Med Biol 1985;186:369–376.

45 Edelson R, Fink J: Le rôle immunitaire de la peau. Pour la Science 1985;94:59–67.

46 Anderson AO, Rubin DH: Effect of avidin on enteric antigen uptake and mucosal immunity to retrovirus. Adv Exp Med Biol 1985;186:579–590.

47 Braquet P, Hosford D, Braquet M, et al: Role of cytokines and platelet-activating factor in microvascular immune injury. Int Arch Allergy Appl Immunol 1989;88: 88–100.

48 Goldstein RA: Immune reaction in a changing environment. Int Arch Allergy Appl Immunol 1989;88:256–258.
49 Grau GE, Piguet PF, Vassalli P, et al: Involvement of tumor necrosis factor and other cytokines in immune-mediated vascular pathology. Int Arch Allergy Appl Immunol 1989;88:34–39.
50 Gray D: Is the survival of memory B cells dependent on the persistence of antigen? Adv Exp Med Biol 1989;237:203–207.
51 Seijen HG, Bun JCAM, Wubbena AS, et al: The germinal center precursor cell is surface μ- and δ-positive. Adv Exp Med Biol 1989;237:233–237.
52 Vonderheide RH, Hunt SV: Surface Igγ D phenotype of rat germinal centre precursor cells. Adv Exp Med Biol 1989;237:239–243.
53 Kishimoto T: Factors affecting B cell growth and differentiation. Annu Rev Immunol 1985;3:133–157.
54 O'Garra A, Umland S, De France T, et al: 'B cell factors' are pleiotropic. Immunol Today 1988;9:45–54.
55 Kehrl JH, Muraguchi A, Fauci AS: The modulation of membrane Ia on human B lymphocytes. Cell Immunol 1985;92:391–403.
56 Suzuki T, Sanders SK, Butler JL, et al: Identification of an early activation antigen (Bac-1) on human B cells. J Immunol 1986;136:1208.
57 Rousset F, De Waal Malefigt R, Selierendregt B, et al: Regulation of Fc receptor for IgE (Cd23) and class II MHC antigen expression on Burkitt's lymphoma cell lines by human IL4 and IFN gamma. J Immunol 1988;140:2625–2631.
58 Zola H, Barcaly S, Furness V, et al: B lymphocyte/carcinoma antigen (BLCa) functional study in B cells. Immunol Cell Biol 1988;66:199–209.
59 Kikutani H, Suemara M, Owaki H, et al: $F_c\varepsilon$ receptor, a specific differentiation marker transiently expressed on mature B cells before isotype switching. J Exp Med 1986;164:1455–1469.
60 Delespesse S, Sarfati M, Hofstetten H: Human Ig-E binding factors. Immunol Today 1989;10:159–164.
61 Kraal G, Twisk A: Interaction of high endothelial venules with T and B cells after antigenic stimulation. Adv Exp Med Biol 1985;186:609–614.
62 Kraal G, Hardy RR, Gallatin WM, et al: Antigen-induced changes in B cell subsets in lymph nodes: Analysis by dual fluorescence flow cytofluorometry. Eur J Immunol 1986;16:829–834.
63 Hendricks HR, Eestersmans IL: Disappearance and reappearance of high endothelial venules and immigrating lymphocytes in lymph nodes deprived of afferent lymphatic vessels: A possible regulatory role of macrophages in lymphocyte immigration. Eur J Immunol 1983;13:663–669.
64 Steinman RM, Inaba K: The binding of antigen presenting cells to T lymphocytes. Adv Exp Med Biol 1989;237:31–41.
65 Sinclair NRSC, Ganoskaltsis A: The immunoregulatory apparatus and autoimmunity. Immunol Today 1988;9:260–263.
66 Gray D: Recruitment of virgin B cells into an immune response is restricted to activation outside lymph follicles. Immunology 1988;65:73–79.
67 Kroese FGM, Wubbena AS, Seijen H, et al: Germinal centers develop oligoclonally. Eur J Immunol 1987;17:1069–1072.
68 Kocks C, Rajewsky K: Stable expression and somatic hypermutation of antibody V regions in B-cell developmental pathways. Annu Rev Immunol 1989;7:537–539.

69 Rizvi N, Chaturvedi UC, Mathur A: Obligatory role of macrophages in dengue virus antigen presentation to B lymphocytes. Immunology 1989;67:38–43.
70 Reichert RA, Gallatin WM, Weissman LL, et al: Germinal center B cells lack homing receptors necessary for normal lymphocyte recirculation. J Exp Med 1984;157:813–827.
71 Fliedner TM: On the origin of tingible bodies in germinal centers; in Cottier H, Odartchenko N, Schindler R, Congdon CC (eds): Germinal Centers in Immune Responses. Berlin, Springer, 1967, pp 218–224.
72 Odartchenko N, Lewerenz M, Sordat B, et al: Kinetics of cellular death in germinal centers of mouse spleen; in Cottier H, Odartchenko N, Schindler R, Congdon CC (eds): Germinal Centers in Immune Responses. Berlin, Springer, 1967, pp 212–217.
73 Heusermann U, Zurborn KH, Schroeder L, et al: The origin of the dendritic reticulum cell: An experimental enzyme-histochemical and electron microscopic study on the rabbit spleen. Cell Tissue Res 1980;209:279–294.
74 Dijkstra CD, Kamperdijk EWA, Döpp EA: The ontogenetic development of the follicular dendritic cells: An ultrastructural study by means of intravenously injected horseradish peroxidase (HRP)-anti HRP complexes as marker. Cell Tissue Res 1984; 236:203–206.
75 Humphrey JH, Grennan D, Sundaram V: The origin of follicular dendritic cells in the mouse and the mechanism of trapping of immune complexes on them. Eur J Immunol 1984;14:859–863.
76 Schriever F, Freddman AS, Freeman G, et al: Isolated human follicular dendritic cells display a unique antigenic phenotype. J Exp Med 1989;169:2043–2058.
77 Jeurissen P, Dijkstra CD: Characteristics and functional aspects of nonlymphoid cells in rat germinal centers, recognized by two monoclonal antibodies ED5 and ED6. Eur J Immunol 1986;16:562–568.
78 Kroese GFM, Wubbena AS, Kuijpers KC, et al: The ontogeny of germinal centre forming capacity of neonatal rat spleen. Immunology 1987;60:597–602.
79 Szakal AK, Hanna MG: The ultrastructure of antigen localization and virus like particles in mouse spleen germinal centers. Exp Mol Pathol 1968;8:75–83.
80 Terashima K, Ikai K, Tajima K, et al: Morphological diversity of DRC-1 positive cells: Human follicular dendritic cells and their relatives. Adv Exp Med Biol 1989; 237:157–163.
81 Chen LI, Frank AM, Adams JC, et al: Distribution of horseradish peroxidase (HRP-) anti-HRP immune complexes in mouse spleen with special reference to follicular dendritic cells. J Cell Biol 1978;79:184–199.
82 Kinet-Denoël C, Heinen E, Radoux D, et al: Follicular dendritic cells in lymph nodes after X-irradiation. Int J Radiat Biol 1982;42:121–130.
83 Nossal GJV, Abbot A, Mitchell J, et al: Antigens in immunity: XV. Ultrastructural features of antigen capture in primary and secondary lymphoid follicles. J Exp Med 1968;127:277–289.
84 Herd ZL, Ada GL: The retention of ^{125}I-immunoglobulins: IgG subunits and antigen-antibody complexes in rat footpads and draining lymph. Aust J Exp Biol Med Sci 1969;47:63–72.
85 Papamichail M, Gutierrez C, Embling P, et al: Complement dependency of localization of aggregated IgG in germinal centers. Scand J Immunol 1975;4:343–347.
86 Klaus GGB, Humphrey JH, Kunkl A, et al: The follicular dendritic cell: Its role in antigen presentation in the generation of immunological memory. Immunol Rev 1980;53:3–28.

87 Reynes M, Aubert JP, Cohen JH, et al: Human follicular dendritic cells express CR_1, CR_2 and CR_3 complement receptor antigens. J Immunol 1985;135:2687–2694.

88 Radoux D, Heinen E, Kinet-Denoël C, et al: Precise localization of antigens in follicular dendritic cells. Cell Tissue Res 1984;235:267–274.

89 Heinen E, Radoux D, Kinet-Denoël C, et al: Isolation of follicular dendritic cells from human tonsils and adenoids: II. Analysis of their Fc receptors. Immunology 1985;54:777–784.

90 Hanna G, Szakal AK: Localization of ^{125}I-labelled antigen in germinal centers of mouse spleen: Histologic and ultrastructural autoradiographic studies of the secondary immune reaction. J Immunol 1968;101:949–962.

91 Tew JG, Mandel TE, Burgess AW: Retention of intact HSa for prolonged periods in the popliteal lymph nodes of specifically immunized mice. Cell Immunol 1979;45:207–212.

92 Tew JG, Mandel T: The maintenance and regulation of serum antibody levels: Evidence indicating a role for antigen retained in lymphoid follicles. J Immunol 1978;120:1063–1069.

93 Heinen E, Radoux D, Kinet-Denoël C, et al: Colloidal gold, a useful marker for antigen localization on follicular dendritic cells. J Immunol Methods 1983;59:361–368.

94 Tenner-Racz K, Racz P, Gartner S, et al: A typical virus particle in HIV-1 associated persistent generalized lymphadenopathy. Lancet 1988;i:774–775.

95 Amstrong JA: Retrovirus localization and ultrastructural pathology in AIDS. Philipps Anal 1989;125:21–23.

96 Kamperdijk EWA, Döpp EA, Dijkstra CD: Transport of immune complexes from the subcapsular sinus into the lymph node follicles of the rat. Adv Exp Med Biol 1989;237:191–196.

97 van Rooijen N, Kors N: Mechanism of follicular trapping: Double immunocytochemical evidence for a contribution of locally produced antibodies in follicular trapping of immune complexes. Immunology 1985;55:31–34.

98 Brown JC, Harris G, Papamichail M, et al: The localization of aggregated human gamma globulin in the spleens of normal mice. Immunology 1973;24:955–968.

99 Radzun HJ, Parwaresh MR: Differential immunohistochemical resolution of the human mononuclear phagocyte system. Cell Immunol 1983;82:174–183.

100 Szakal AK, Holmes KL, Tew JG: Transport of immune complexes from the subcapsular sinus to lymph node follicles on the surface of non phagocytic cells, including cells with dendritic morphology. J Immunol 1983;131:1714–1727.

101 Gray D, Kumaratne DS, Lortan J, et al: Relation of intrasplenic migration of marginal zone B cells to antigen localization on follicular dendritic cells. Immunology 1984;52:659–669.

102 Heinen E, Braun M, Coulie PG, et al: Transfer of immune complexes from lymphocytes to follicular dendritic cells. Eur J Immunol 1986;16:167–172.

103 Braun M, Heinen E, Cormann N, et al: Influence of immunoglobulin isotypes and lymphoid cell phenotype on the transfer of immune complexes to follicular dendritic cells. Cell Immunol 1987;107:99–106.

104 Pallesen G, Myrhe-Jensen O: Immunophenotypic analysis of neoplastic cells in follicular cell sarcoma. Leukemia 1987;1:549–557.

105 Johnson GD, Hardie DL, Ling NR, et al: Human follicular dendritic cells (DC): A study with monoclonal antibodies (MoAb). Clin Exp Immunol 1986;64:205–213.

106 Heinen E, Kinet-Denoël C, Simar JS: 5-Nucleotidase activity in isolated follicular dendritic cells. Immunol Lett 1985;9:75–80.

107 Heinen E, Kinet-Denoël C, Radoux D, et al: Mouse lymph node follicular dendritic cells, quantitative analysis and isolation. Adv Exp Med Biol 1985;186:171–183.

108 Peters JPJ, Radermakers LHPM, Roelopp JMM, et al: Distribution of dendritic reticulum cells in follicular lymphoma and reactive hyperlasia: Light microscopic identification and general morphology. Virchows Arch [B] 1984;46:215–228.

109 Lilet-Leclercq C, Radoux D, Heinen E, et al: Isolation of follicular dendritic cells from human tonsils and adenoids: I. Procedures and morphological characterization. J Immunol Methods 1984;66:235–244.

110 Heinen E, Lilet-Leclercq C, Mason D, et al: Isolation of follicular dendritic cells from human tonsils or adenoids: II. Immunocytochemical characterization. Eur J Immunol 1984;14:267–273.

111 Tsunoda R, Kojima M: A light microscopical study of isolated follicular dendritic cell-clusters in human tonsils. Acta Pathol Jpn 1987;37:575–585.

112 Cormann N, Lesage F, Heinen E, et al: Isolation of follicular dendritic cells from human tonsils and adenoids: V. Effect on lymphocyte proliferation and differentiation. Immunol Lett 1986;14:29–35.

113 Tsunoda R, Cormann N, Heinen E, et al: Cytokines produced in lymph follicles. Immunol Lett 1989;22:119–134.

114 Lortan J, Roobottorm CA, Oldfield S, et al: Newly produced virgin B cells migrate to secondary lymphoid organs but their capacity to enter follicles is restricted. Eur J Immunol 1987;17:1311–1316.

115 Mills GC, Schmalsteig FC, Trimmer KB, et al: Purine metabolim in adenosine deaminase deficiency. Proc Natl Acad Sci USA 1976;73:2867.

116 Ferraris PC, Coffey RG, Hadden JW: Substituted purines as immunomodulators: Speculative considerations. EOS Rev Immunol Immunolofarma 1984;4:134–141.

117 Awad AB, Chattopadhyay JP: Developmental alterations in 5'-nucleotidase kinetics and lipid composition of rat heart sarcolemma. Mech Ageing Dev 1984;22:151–158.

118 Ross GD, Atkinson JP: Complement receptor structure and function. Immunol Today 1985;6:115–119.

119 Gutman G, Weissman I: Lymphoid tissue architecture: Experimental analysis of the origin and distribution of T and B cells. Immunology 1972;23:465.

120 Roux RV, Ledbetter JA, Weissman I: Mouse lymph node germinal centers contain a selected subset of T cells: The helper phenotype. J Immunol 1982;128:2243–2247.

121 Yamanaka N, Sambe S, Harabuchi Y, et al: Immunohistological study of tonsil distribution of T cell subsets. Acta Otolaryngol 1983;96:509–516.

122 Kroese FGM: The Generation of Germinal Centers; thesis, Groningen, 1987.

123 Katz P, Whalen G, Cupps TR, et al: Natural killer cells can enhance the proliferative responses of B lymphocytes. Cell Immunol 1989;120:270–276.

124 Velardi A, Mingari MC, Moretta L, et al: Functional analysis of cloned germinal center CD4+ cells with natural killer cell-related features: Divergence from typical T helper cells. J Immunol 1986;137:2808–2813.

125 Wieczorek R, Jakobiec FA, Sacks EH, et al: The immunoarchitecture of the normal human lacrimal gland: Relevancy for understanding pathologic conditions. Ophthalmology 1988;95:100–109.

126 Jacobson EB, Caporale LH, Thorbecke GJ: Effect of thymus cell injections on

germinal center formation in lymphoid tissues of nude mice. Cell Immunol 1974; 13:416–430.

127 Daeron M, Neauport-Sautès C, Yodoi J, et al: Receptors for immunoglobin isotypes (FcR) on murine T cells: II. Multiple FcR induction on hybridoma T cell clones. Eur J Immunol 1985;15:668–674.

128 Klatzmann D, Gluckman JC: HIV infection: Facts and hypotheses. Immunology 1986;7:291–296.

129 Kotani M, Okada K, Fuji H, et al: Lymph macrophages enter the germinal center of lymph nodes of guinea pigs. Acta Anat 1977;99:391–402.

130 Rademakers LHPM, De Weger RA, Roholl PJM: Identification of alkaline phosphatase positive cells in human germinal centres as follicular dendritic cells. Adv Exp Med Biol 1989;237:165–168.

131 Mollo F, Monga G, Stramignoni A: Dark reticular cells in human lymph adenitis and lymphoma. Virchows Arch [B] 1969;3:117–126.

132 Jalkanen S, Jalkanen M, Bargatze R, et al: Biochemical properties of glycoproteins involved in lymphocyte recognition of high endothelial venules in man. J Immunol 1988;141:1615–1623.

133 Coombe R, Rider CC: Lymphocyte homing receptors cloned: A role for anionic polysaccharides in lymphocyte adhesion. Immunol Today 1988;10:289–291.

134 Heinen E, Cormann N, Braun M, et al: Isolation of follicular dendritic cells from human tonsils and adenoids: VI. Analysis of prostaglandin secretion. Ann Inst Pasteur Immunol 1987;137:369–382.

135 Bosseloir A, Hooghe-Peters EL, Heinen E, et al: Localization of interleukin 6 (IL6) mRNA in human tonsils by in situ hybridization. Eur J Immunol 1989;19:2379–2381.

136 Lasky JL, Thorbecke GJ: Growth requirement of SJL lymphomas in vitro: Effect of BCGF II. Adv Exp Med Biol 1989;237:145–151.

137 Scher I, Sharrow SO, Paul WE: X-linked B-lymphocyte defect in CBA/N mice: II. Abnormal development of B-lymphocyte populations defined by their density of surface immunoglobulin. J Exp Med 1977;144:507–519.

138 Hayakawa K, Hardy RR, Honda M, et al: Ly-1 cells: Functionally distinct lymphocytes that secrete IgM autoantibodies. Proc Natl Acad Sci USA 1984;81:2494–2498.

139 Herzenberg L, Lalor P, Stall AM, et al: Characteristics of Ly-1 B cells; in Cinader B, Miller RG (eds): Progress in Immunology. Orlando, Academic Press, 1986, pp 376–384.

140 Matsuno K, Ezaki T, Kotani M: Localization of T-I2 antigen (Ficoll-FITC) and corresponding specific anti-FITC antibody forming cells in the rat spleen. Adv Exp Med Biol 1989;237:895–899.

141 Yong-Jun L, Oldfield S, Mac Lennan ICM: Thymus-independent type 2 responses in lymph nodes. Adv Exp Med Biol 1989;237:113–117.

142 Sablitzky F, Wildner G, Rajewsky K: Somatic mutation and clonal expansion of B cells in an antigen driven immune response. EMBO J 1985;4:345–350.

143 Rajewski K: Somatic antibody mutants. Immunol Today 1989;10:S10–S11.

144 Kraal G, Weissman IL, Butcher E: Germinal centre B cells: Antigen specificity and changes in heavy chain class expression. Nature 1982;298:377–379.

145 Rajewsky K, Forster I, Cumano A: Evolutionary and somatic selection of the antibody repertoire in the mouse. Science 1987;238:1088–1094.

146 Fridman WH, Rabourdin-Combe C, Néauport-Sautès C, et al: Characterization and function of T cell Fcγ receptor. Immunol Rev 1981;56:51–73.

147 Paul WE, Mizuguchi J, Brown M, et al: Regulation of B-lymphocyte activation, proliferation and immunoglobulin secretion. Cell Immunol 1986;99:7–13.

148 Millet I, Samarut C, Revillard JP: Class-specific suppression of human B cell maturation by IgA-binding factors. Eur J Immunol 1988;18:545–550.

149 Harriman GR, Kunimoto DY, Elliott JF, et al: The role of IL5 in IgA B cell differentiation. J Immunol 1988;140:3033–3039.

150 Iskizaka K, Yodoi J, Suemura M, et al: Isotype-specific regulation of the IgE-binding factors. Immunol Today 1983;4:192–196.

151 Durkin HG, Chice SM, Jiang SI: Origin and fate of IgE-bearing lymphocytes: Modulation of IgE isotype expression on Peyer's patch cells. Adv Exp Med Biol 1985;186:127–130.

152 Ishizaka K: Regulation of the IgE antibody response. Int Arch Allergy Appl Immunol 1989;88:8–13.

153 Vercelli D, Geha RS: Regulation of IgE synthesis in humans. J Clin Immunol 1989;9:75–83.

154 Coulie PG, Van Snick J: Enhancement of IgG anti-carrier responses by IgG_2 anti-hapten antibodies in mice. Eur J Immunol 1985;15:793–797.

155 Dougherty GJ, Selvendran Y, Murdock S, et al: The human mononuclear phagocyte high-affinity Fc receptor, FcRI, defined by a monoclonal antibody, 10.1. Eur J Immunol 1987;17:1453–1459.

156 Butcher EC: Lymphocyte migration and mucosal immunity; in Heyworth MF, Jones AL (eds): Immunology of the Gastrointestinal Tract and Liver. New York, Raven Press, 1988, pp 93–103.

157 Pals ST, den Otter A, Miedema F, et al: Mel 14 and LFA-1 organ specific and non-specific adhesion molecules involved in homing of human lymphocytes. Adv Exp Med Biol 1989;237:505–509.

158 Holzmann B, McIntyre BW, Weissman IL: Identification of a murine Peyer's patch specific lymphocyte homing receptor as an integrin molecule with an α chain homologous to human VLA-4α. Cell 1989;56:37–46.

159 Caligaris-Cappio F, Bergui L, Tesio, et al: Interrelationships of cell-substrate and cell-cell adhesion structures in B-chronic lymphocytic leukemia (B-CLL) cells. Adv Exp Med Biol 1989;237:133–137.

160 Coico RF, Bhogal BS, Thorbecke GJ: Relationship of germinal centers in lymphoid tissue to immunologic memory: VI. Transfer of B cell memory with lymph node cells fractionated according to their receptors for peanut agglutinin. J Immunol 1983;131:2254–2257.

161 Mandel TE, Phipps RP, Abbot A, et al: The follicular dendritic cell: Long-term antigen retention during immunity. Immunol Rev 1980;53:30–59.

162 Kölsch E, Oberbarnscheidt J, Brüner K, et al: The Fc-receptor: Its role in the transmission of differentiation signals. Immunol Rev 1983;49:61–78.

163 Uher F, Lamers MC, Dikler HB: Antigen-antibody complexes bound to B-lymphocyte Fcγ receptors regulate B-lymphocyte differentiation. Cell Immunol 1985;95:368–379.

164 Oldfield S, Yon-Jun L, Beaman M, et al: Memory B cells generated in T cell-dependent antibody responses colonise the splenic marginal zone. Adv Exp Med Biol 1989;237:93–98.

165 Dobashi M, Tersashima K, Imai Y: Electron microscopic study of differentiation of antibody-producing cells in mouse lymph nodes after immunization with horseradish peroxidase. J Histochem Cytochem 1982;30:67–74.

166 Latron F, Jotterand-Bellomo M, Maffei A, et al: Active suppression of major histocompatibility complex class II gene expression during differentiation from B cells to plasma cells. Proc Natl Acad Sci USA 1988;85:2229–2233.

167 Bienenstock J, Dolezel J: Peyer's patches: Lack of specific antibody-containing cells after oral and parenteral immunization. J Immunol 1971;106:938–945.

168 Geldof AA, Van den Ende M, Janse EM, et al: Specific antibody formation in mouse spleen: Histology and kinetics of the secondary immune response against HRP. Cell Tissue Res 1983;231:135–142.

169 Van Rooijen N, Kors N, Van Nieuwmegen R: The development of specific antibody producing cells in the spleen of rabbits during the primary and secondary immune response. Adv Exp Med Biol 1985;186:153–159.

170 Tew JG, Burton GF, Szaka A, et al: A subpopulation of germinal center B cells differentiate directly into antibody forming cells upon secondary immunization. Adv Exp Med Biol 1989;237:215–220.

171 Grobler P, Buerki H, Cottier H, et al: Cellular bases for relative radioresistance of the antibody-forming system at advanced stages of the secondary response to tetanus toxoid in mice. J Immunol 1974;112:2154–2165.

172 Yoshizaki K, Matusda T, Nishimoto N, et al: Pathogenic significance of interleukin 6 in Castleman's disease. Blood, in press.

173 Zola H, Nikoloutsopoulos A: Effect of recombinant human tumor necrosis factor beta (TNF) on activation, proliferation and differentiation of human B lymphocytes. Immunology 1989;67:231–236.

174 Berkovitz BKB, Holland GR, Moscham BJ: A Colour Atlas and Text Book of Oral Anatomy. London, Wolfe, 1978.

175 Husband AJ: Kinetics of extravasation and redistribution of IgA-specific antibody-containing cells in the intestine. J Immunol 1982;128:1355–1359.

176 Czinn SJ, Lamm ME: Selective chemotaxis of subsets of B lymphocytes from gut-associated lymphoid tissue and its application for the recruitment of mucosal plasma cells. J Immunol 1989;136:3607–3611.

177 Heinen E, Defresne MP, Boniver J, et al: Les organes du système immun; in Pastoret PP, Silim A, Govaerts A, Bazin H (eds): Immunologie animale. Paris, Flammarion, 1990, pp 51–80.

178 Heyworth MF: T cells and other non-B lymphocytes; in Heyworth MF, Jones AL (eds): Immunology of the Gastrointestinal Tract and Liver. New York, Raven Press, 1988, pp 1–22.

179 Kawaniski H, Saltzman LE, Strober W: Mechanisms regulating IgA class-specific immunoglobulin production in murine gut-associated lymphoid tissues: I. T cells derived from Peyer's patches that switch sIgM B cells to sIgA cells in vitro. J Exp Med 1983;157:433–450.

180 Beagley KW, Elridge JH, Kiyono H, et al: The identification of murine Peyer's patches T cell-derived factors which enhance IgA responses. Adv Exp Med Biol 1989;237:641–647.

181 Staite ND, Panayi GS: Prostaglandin regulation of lymphocyte-B function. Immunol Today 1984;5:175–177.

182 Hijmans W, Schuit HRE, Hulsing-Hesselink E: An immunofluorescence study on intracellular Ig in human bone marrow cells. Ann NY Acad Sci 1971;177:290–305.

183 Mac Millan R, Longmire RL, Yelenosky RJ, et al: Bone marrow as a major site of human IgG production. Blood 1972;40:926.

184 Benner R, Van Oudeharen A, Björklund M, et al: Background immunoglobulin production: Measurement, biological significance and regulation. Immunol Today 1982;3:243–249.

185 Geldof AA, Rijnhart P, van den Ende M, et al: Morphology, kinetics and secretory activity of antibody-forming cells. Immunobiology 1984;166:296–307.

186 Ohno T, Miyama-Inaba M, Masuda T, et al: Inhibitory mechanism of the proliferative responses of resting B cells: Feedback regulation by a lymphokine (suppressive B-cell factor) produced by Fc receptor-stimulated B cells. Immunology 1987;61:35–42.

187 Ballas ZH, Feldbush TL, Needlman W, et al: Complement inhibits immune responses: C_3 preparations inhibit the generation of human cytotoxic lymphocytes. Eur J Immunol 1983;13:279–284.

188 Erdei A, Fust G, Gying J, et al: C_3b acceptors on human peripheral blood mononuclear cells: Characterizations and functional role. Immunology 1983;49:423–430.

189 Meuth JL, Morgan EL, Discipio RG: Suppression of T lymphocyte functions by human C_3 fragments: I. Inhibition of human T cell proliferative responses by a kallikrein cleavage fragment of human iC_3b. Immunology 1983;130:2605–2611.

190 Hashimoto F, Sakiyama Y, Matsumoto S: The suppressive effect of gammaglobulin preparations on in vitro pokeweed mitogen-induced immunoglobulin production. Clin Exp Immunol 1986;65:409–415.

191 Virgin HW, Wittenberg GF, Unanue ER: Immune complex effects on murine macrophages: I. Immune complexes suppress interferon induction of Ia expression. J Immunol 1985;135:3735–3743.

192 Heyman B, Pilström L, Schulmans MJ: Complement activation is required for IgM-mediated enhancement of the antibody response. J Exp Med 1988;167:1999–2004.

193 Lisowska-Grospierre B, Bohler MC, Fischer A, et al: Defective membrane expression of the LFA-1 complex may be secondary to the absence of the β chain in a child with bacterial infection. Eur J Immunol 1986;16:205–208.

194 Giblett ER, Amman AJ, Wara DW, et al: Nucleoside-phosphorylase deficiency in a child with severely defective T-cell immunity and normal B-cell immunity. Lancet 1975;i:1010–1013.

195 Roitt IM, Brostoff J, Male DK: Immunologie fondamentale et appliquée. Oxford, Blackwell, 1985.

196 Armstrong JA, Horne R: Follicular dendritic cells and virus-like particles in AIDS-related lymphadenopathy. Lancet 1984;ii:370–373.

197 Yagawa K, Kalu M, Ichinose Y, et al: Down regulation of the receptor expression in guinea pig peritoneal exudate macrophages by muramyl dipeptide or lipopolysaccharide. J Immunol 1985;134:3705–3711.

198 Heinen E, Cormann N, Kinet-Denoël C, et al: Lipopolysaccharide suppresses immune complex retention by follicular dendritic cells without cytological alterations. Immunol Lett 1986;13:323–327.

199 Stein H, Gerdes J, Mason DY: The normal and malignant germinal centre. Clin Haematol 1982;11:531–559.

200 Louis E, Philippet B, Cardos B, et al: Intercellular contacts between germinal center cells. Mechanisms of adhesion between lymphoid cells and follicular dendritic cells. Acta ORL 1990;43:297–320.

Ernst Heinen, MD, Institute of Human Histology, University of Liège, 20, rue de Pitteurs, B–4020 Liège (Belgium)

Sorg C (ed): Molecular Biology of B Cell Developments.
Cytokines. Basel, Karger, 1990, vol 3, pp 61–84

Immunoglobulin Heavy-Chain Class Switching: Molecular Requirements for Constant-Region Gene Switch Recombination

David E. Ott, Moon G. Kim, Kenneth B. Marcu[1]

Departments of Biochemistry and Cell Biology, Microbiology and Pathology,
State University of New York at Stony Brook, N.Y., USA

Immunoglobulin heavy-chain isotype switch recombination allows for the committed expression of μ-, γ-, ϵ- and α-constant (C) region isotypes. Switch recombination results in the replacement of the $C\mu$ gene segment, $3'$ of a functionally rearranged variable (V) region gene, for 1 of 7 downstream immunoglobulin C_H gene segments. Switch (S) regions are positioned $\sim$2–2.5 kilobases (kb) $5'$ of each C_H gene segment with the exception of $C\delta$. S regions consist in part of short tandemly repetitive sequences which display varying degrees of sequence homology to each other [reviewed in ref. 1–5].

The lengths and sequence organization of tandem repeats vary considerably in different S regions. GAGCT and GGGGT sequences are found in all S regions in addition to 3 other commonly observed pentamers (e.g. ACCAG, GCAGC and TGAGC) [1–5]. A heptamer consensus motif (YAGGTTG, $Y = C$ or T) has also been found nearby the majority of switch recombination sites in plasma cell tumors and hybridoma lines [1]. These sequences are organized into larger tandemly repeated structures in every S region except for $S\mu$. The $S\mu$ region consists of two parts, a $5'$-YAGGTTG consensus region and a $3'$-segment that consists of a virtual solid block of $(GAGCT)_n GGGGT$ sequences [2, 3, 7]. The $S\gamma_{2b}$ region is made up of tandemly repeated 49mer

[1] We thank Dr. Fred Alt for providing us with the 18-8tk⁻, 38B9tk⁻ and 300-18tk⁻ pre-B lines. We also gratefully acknowledge the assistance of Joyce Schirmir and Chris Helmke for figures and photographic work. This research was supported by US NIH grant GM26939.

consensus repeats which are not highly homologous to the Sμ region [8, 9]. DNA sequence analysis of rearranged C_H gene switch sites in plasma cell tumors, B lymphomas, hybridomas and a few A-MuLV-transformed pre-B lines have generally not revealed conserved recognition sequences at their recombination sites, although a YAGGTTG consensus sequence and GAGCT/GGGGT motifs are found nearby most switch sites [1–5].

The existence of novel repeat units in different S regions and the variations in their overall sequence homologies originally led to the suggestion that isotype switching may involve C_H-class-specific recombinases [10]. It could be argued that this idea has gained support from the documented propensity of A-MuLV-derived cell lines to switch from μ to $γ_{2b}$ [11–15], and from the preferential C_H isotype switching in a murine B lymphoma [16, 17] in some myeloma tissue culture variants [18, 19] and in lipopolysaccharide (LPS)- or LPS/BSF-1-stimulated normal splenic B cells [4, 20]. However, it has also been proposed that these apparent instances of directed C_H switching may be predetermined by factors which preferentially enhance the accessibility of different C_H gene segments to a general S region recombinase [21–23]. It is interesting to note in this context that the assembly of immunoglobulin V and C gene segments may also be associated with their unique transcriptional competence or accessibility in B lymphoid cells [23–27].

To directly investigate the molecular requirements and regulation of C_H class switching, we have designed selectable retroviral vectors as in vivo substrates for switch recombination [28, 29]. In an independent though parallel approach toward defining the components of the switching mechanism, we have begun to identify B-cell-specific nuclear factors which specifically interact with S segment sequences.

Materials and Methods

Tissue Culture

The 18-8tk⁻, 38B9tk⁻ and 300-18tk⁻ pre-B [14, 30] and A39Rγ1.1tk⁻ hybridoma [4] cell lines were maintained in RPMI 1640 supplemented with 10% fetal calf serum (FCS), 2 mM glutamine and 50μM β-mercaptoethanol or in Dulbecco's modified essential medium (DME) supplemented with 10% FCS, respectively. Ltk⁻ and NIH3T3tk⁻ cells were grown in DME with 10% calf serum. Recombinant retroviruses were produced by the ψ2 helper-free ecotropic packaging cell line [28, 31, 32] except for Ltk⁻ cell infections which required the amphotypic capsid provided by the PA12 packaging line [33]. Retroviral infections and drug selections were carried out as described [28]. The A39Rγ1.1 cell line (gift of Sigrid Klein) was made thymidine kinase (tk)-deficient by exposing 5×10^6 cells to a dose of germicidal ultraviolet light sufficient to yield about 60%

cell mortality. Surviving cells were grown to a density of 5×10^6 and then placed in a 24-well Costar plate with 100 μM 5-bromodeoxyuridine (BrdU)-supplemented media to select for thymidine-kinase-deficient mutants, which appeared within 3 weeks. These tk⁻ cell lines were expanded for about 1 week without selection before separate platings in 100 μM BrdU or HAT media. Genuine thymidine kinase mutants were resistant to BrdU but sensitive to HAT. Fluctuation frequency analysis was carried out on 5–7 independent populations of ZN(Sμ/Sγ$_{2b}$)tkl-infected cells as previously described [28, 34].

Isolation of Recombinant Proviruses

Recombined ZN(Sμ/Sγ$_{2b}$)tkl viruses were cloned from previously characterized 18-8tk⁻ clones by two different methods. The first involved the fusion of the 18-8tk⁻ clones with COS cells by mixing 1×10^7 18-8tk⁻ with 5×10^6 COS cells in the presence of 48% PEG-1000 in DME followed by centrifugation in a Beckman model B microfuge for 20 s. The fusion mixture was diluted with 10 ml of DME, centrifuged for 5 min at 750 rpm at 4 °C in a Sorval RT6000 centrifuge, resuspended in 10 ml of DME/10% FCS and placed in a 10-cm plate. After 3 days the cells were split into 0.5 mg/ml G-418-containing medium to select for fused cells, which would be adherent and resistant to G-418. Surviving fused cells were expanded up to 5×10^7 and Hirt lysates were prepared to isolate circular proviral LTR-LTR recombinants which had been excised from the cellular genome. By this strategy, circular-ized proviruses were amplified due to the presence of an SV40 replication origin in the ZN(Sμ/Sγ$_{2b}$)tkl vector, and the necessary DNA replication factors provided by the COS fusion partner. Hirt lysates were used to transform DH5 competent cells (Bethesda Research Labs) followed by selection in kanamycin plates. The second method was plasmid rescue. Genomic DNAs from clones of G-418/BrdU-resistant, ZN(Sμ/Sγ$_{2b}$)tkl-infected 18-8tk⁻ cells were restricted with SstI, which cleaves once in each LTR and once in the Sμ segment. DNA fragments were purified on a 1% agarose gel, ligated under dilute conditions (4 μg/ml) and used to transform high-efficiency DH5 cells (Bethesda Research Labs) to kanamycin resistance. Clones rescued by either method were restriction-mapped and compared to their original genomic contexts. Proviral clones rescued in this manner were subcloned into either mp18 or mp19 vectors and their inserts sequenced with the Sequenase system (United States Biochemicals).

Southern and Northern Blotting

Southern and Northern blotting and hybridizations of blotted Nytran filters (Schleicher & Schuell) to nick-translated fragments [35] or uniform labeled single-stranded DNAs in M13 vectors [36] were performed as described [37].

DNA-Protein Mobility Shifts

Gel retardation assays were carried out as described by Singh et al. [38] with slight modifications. Crude nuclear extracts were prepared according to Dignam et al. [39] and a standard binding reaction contained 1 ng of end-labeled DNA probe, 5 μg of extract and 4 μg of poly-dA-dT-nonspecific competitor which were incubated for 20 min at room temperature. DNA-protein complexes were separated from unbound probe on 4% poly-acrylamide gels as described [38].

Immunofluorescence

Heavy-chain immunoglobulin expression in 300-18tk⁻ cells was analyzed by immu-nofluorescence with either fluorescein isothiocyanate (FITC)-conjugated goat antimouse μ antibodies or tetramethyl rhodamine (TRIC)-conjugated goat antimouse γ_{2b} antibody (Southern Biotechnology). Cells were fixed, stained as described [17] and then examined under a Leitz fluorescence microscope.

Results

Properties of the ZN(Sμ/Sγ₂ᵦ)tk1 Retrovirus Vector

The derivation of the ZN(Sμ/Sγ₂ᵦ)tk1 retrovirus vector is displayed in figure 1. We chose to design a retroviral vector for monitoring the presence of

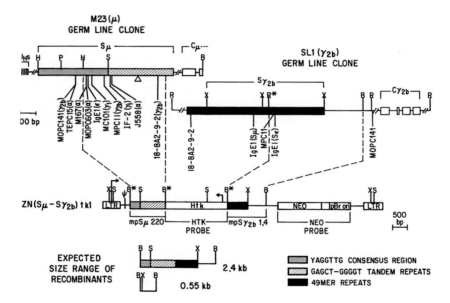

Fig. 1. Derivation and construction of ZN(Sμ/Sγ₂ᵦ)tk1. The two germ line genomic clones (M23 and SL1) that provided the respective Sμ and Sγ₂ᵦ switch region segments are shown at the top [28]. The general sequence composition of the regions is shown by the various shaded boxes. Switch sites used by various mouse myelomas are indicated under the genomic clone maps [28]. An internal deletion in Sμ occurred during clone propagation and is shown as an open triangle. The map of the ZN(Sμ/Sγ₂ᵦ)tk1 construct shows the packaged virus genome and the origin of the chosen S region segments. The probes used are denoted by the parentheses under the maps. The predicted maps of recombinants that utilize the repetitive regions are shown along with their predicted sizes. Enzyme abbreviations are: B = BamHI; B* = BamHI by linker modification; H = HindIII; R = EcoRI; R* = EcoRI site added by cloning procedure; S = SstI; X = XbaI.

switch recombinase activities in B cells for a number of reasons. First, as a retrovirus, it can be easily introduced into B cells, which are normally refractory to most DNA transfection techniques, with high efficiency. More importantly, the retrovirus integrates in a single copy per cell [40], thereby allowing recombinations to be scored as unique events. Moreover, since retroviruses have a defined molecular mechanism of integration, constructs are integrated without perturbation of the sequences residing between the LTRs [40]. The two independent markers (*neo* and HTK) allow for both the maintenance and the selection of proviruses which have undergone S-region-mediated HTK gene deletion. The proximity of the LTR's promoter and enhancer sequences to the HTK gene and the two S regions would ensure that this portion of the vector is accessible for S-region-mediated recombination. Therefore, the intact HTK gene would be transcriptionally active in all NEO-expressing G-418-resistant transformants, thereby obviating the necessity for prior selection in HAT media [27]. Fusion of the $S\mu$ and $S\gamma_{2b}$ sequences would result in HTK gene deletion which is selected, in a thymidine-kinase-deficient cell line, by resistance to BrdU. The novel sizes of the retroviral S region restriction fragments allows them to be readily distinguished from the corresponding endogenous S regions. Recombinants are easily identified by BamHI digestion which releases the $S\mu$, HTK and $S\gamma_{2b}$ sequences of $ZN(S\mu/S\gamma_{2b})tk1$ on differently sized fragments. The $S\mu$ and the $S\gamma_{2b}$ switch regions possess a limited amount of sequence homology [9, 41] which should minimize the background of general homologous recombination.

HTK Phenotype Loss in $ZN(S\mu/S\gamma_{2b})tk1$-Infected Thymidine-Kinase-Negative Pre-B Lines is Mediated by Switch Recombination

Three thymidine-kinase-negative A-MuLV-transformed murine pre-B cell lines ($18\text{-}8tk^-$, $38B9tk^-$ and $300\text{-}18tk^-$) were infected with $ZN(S\mu/S\gamma_{2b})tk1$ virus and then selected for resistance to G-418. The 18-8 line has been documented to spontaneously switch from $C\mu$ to $C\gamma_{2b}$ expression by S-region-mediated deletion [13, 14]. The 38B9 line has DJ rearrangements on both heavy-chain alleles [30] and has been shown to recombine D-J gene segments of both IgH and T cell receptor genes in plasmid vectors at a high frequency [26, 27, 42]. The relative frequencies of HTK phenotype loss were determined by selection in medium supplemented with G-418 and BrdU (see table 1). The 300-18 line was the most efficient in eliminating the HTK phenotype.

To determine the mechanisms of HTK phenotype loss, G-418/BrdU-resistant clones from the frequency analysis and limiting dilution cloning

Table 1. Fluctuation frequency analysis of HTK phenotype loss in ZN((Sμ/Sγ$_{2b}$)tk1-infected cell lines assayed by G-418/BrdU selection

Cell lines containing ZN(Sμ/Sγ$_{2b}$)tk1	HTK phenotype loss/cell/generation[a]	HTK phenotype loss relative to 300-18 cell line
300-18tk⁻	4×10^{-4}	1
18-8tk⁻ᵇ	5×10^{-5}	1.2×10^{-1}
38B9tk⁻	1×10^{-5}	2.5×10^{-2}
A39Rγ 1.1tk⁻	3×10^{-7}	7.5×10^{-4}
Ltk⁻	4×10^{-8}	1.0×10^{-4}
NIH3T3tk⁻	5×10^{-8}	1.2×10^{-4}

[a]Frequency values were derived by the graphical solution of the fluctuation equation as described by Luria and Delbruck [34] applied to multiple cell population pools as described in Ott et al. [28]. [b]Frequency of HTK loss in 18-8tk⁻ cells may be reduced due to diminished viability of 18-8tk⁻ cells [13, 27].

were expanded, and genomic DNAs were prepared and submitted to Southern blot analysis with *neo, Htk* and S region probes. As expected, each of these clones contained a *neo* gene but none retained an *Htk* gene (fig. 2a). All of the G-418/BrdU-resistant 18-8tk⁻ clones displayed a novel Sγ$_{2b}$-hybridizing BamHI fragment (fig. 2b). The identical bands were also positive with an S-specific probe (fig. 2b). Therefore, the novel Sμ/Sγ$_{2b}$-containing BamHI fragment in each independent clone was generated by recombination between the Sμ and Sγ$_{2b}$ sequences which deleted the *Htk* gene and its bordering 5′ and 3′ BamHI sites. The potential size range of the S region recombination products resulting in complete *Htk* gene deletion is shown in figure 2. Recombination between the 3′-end of Sμ and the 5′-most portion of Sγ$_{2b}$ would produce the largest expected fragment size of 2.4 kb, while fusion of the 5′-most portion of Sμ with the 3′-most repeat of Sγ$_{2b}$ would produce the smallest expected fragment size of 0.55 kb. Fourteen independent 18-8tk⁻ clones were analyzed and the structures of all the recombined S regions are displayed in figure 3. S region recombination products ranged in size from 0.80 to 2.0 kb with a median size of 1.3 kb with no 2 clones having the same size. Similar results were obtained with genomic DNAs of G-418/BrdU-resistant clones of 38B9tk⁻ cells and 300-18tk⁻ cells (see maps for representative 38B9 clones in figure 3). However, the higher frequency of HTK phenotype loss in the 300-18tk⁻ line permitted the detection of multiple, rearranged Sμ/Sγ$_{2b}$ co-hybridizing bands in G-418-resistant populations of cells prior to BrdU selection (fig. 4).

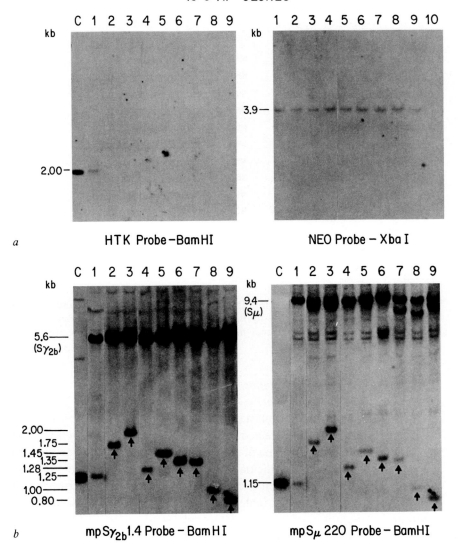

Fig. 2. Southern blot analysis of DNAs from ZN (Sμ/Sγ$_{2b}$tk1-containing 19–8tk⁻ G-418/Brd4-resistant clones. *a* Southern blots of BamHI- and XbaI-digested genomic DNAs probed with HTK or NEO probes, respectively. *b* BamHI digstions probed with Sγ$_{2b}$ and Sμ probes, respectively. Lane C = pZN(Sμ/Sγ$_{2b}$)tk1 vector control; lane 1 = ZN (Sμ/Sγ$_{2b}$)tk1 in 18–8tk⁻ cells before G-418/BrdU selection; lanes 2–9 = G-418/BrdU resistant clones NB32, NB31, NB30, NB26, NB5, NB1, NB3 and NB2, respectively. Bands co-hybridizing with both probes are denoted by arrows. The endogenous S sequences are present above the construct bands.

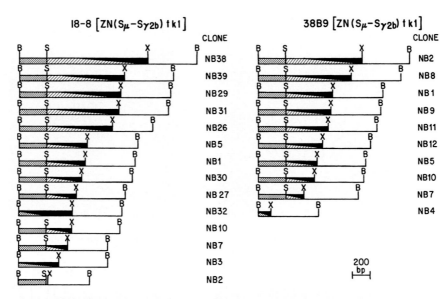

Fig. 3. Southern blot maps of the ZN(Sμ/Sγ2b)tk1-containing 18-8tk⁻ and 38B9tk⁻ clones. Clone DNAs were mapped by BamHI, SstI and XbaI digestion. Recombined inserts are displayed with a diagonal line showing the approximate area where the recombinations have taken place.

Absence of Switch Recombinase Activity in Fibroblast Lines and in a Hybridoma

Ltk⁻ and NIH3T3tk⁻ fibroblast cells and a tk⁻ derivative of the A39Rγ1.1 hybridoma line [4] were infected with the ZN(Sμ/Sγ2b) retrovirus. The A39γ1.1 cell line is a hybridoma of a C57Bl/6 splenic B lymphocyte fused to the immunoglobulin-nonproducing AGX63 mouse myeloma line [4]. Subclones of this cell line have been documented to switch their expression from IgγG1 to IgGγ2b with a variable low frequency [4; Sigrid Klein, pers. commun.]. Populations of G-418-resistant cells were scored for HTK phenotype loss in G-418/BrdU media and individual drug-resistant clones were examined by Southern blotting. As shown in table 1, their frequencies of HTK phenotype loss are 1,000- to 10,000-fold lower than observed for the 300-18 line. Retrovector *Htk* genes were retained in all G-418/BrdU-resistant, infected clones and there was no evidence of S segment rearrangements [28, 29]. The lack of S region recombination to achieve HTK gene inactivation in fibroblasts suggests that S-region-mediated recombination within the retroviral vector is B-cell-specific and *Htk* gene inactivation is likely to occur

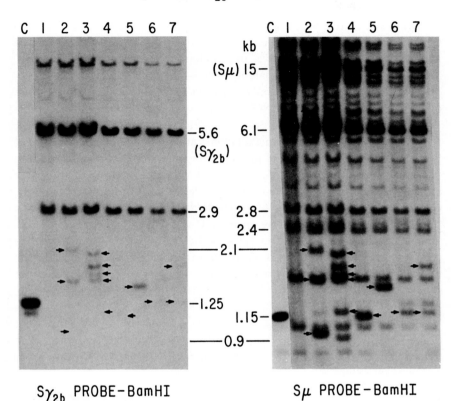

$300\text{-}18\left[ZN(S\mu/S\gamma_{2b})\,tk1\right]$ POPULATIONS

$S\gamma_{2b}$ PROBE-BamHI $S\mu$ PROBE-BamHI

Fig. 4. Southern blot analysis of BamHI-digested DNAs from $ZN(S\mu/S\gamma_{2b})tk1$-infected 300-18tk⁻ G-418-resistant populations prior to BrdU selection. Blots were hybridized to either $S\gamma_{2b}$ or $S\mu$ probes. Lane C = $pZN(S\mu/S\gamma_{2b})tk1$ vector control; lane 1 = uninfected 300-18 tk⁻; lanes 2–7 = $ZN(S\mu/S\gamma_{2b})tk1$-infected 300-18tk⁻ G-418-resistant populations Nos. 1, 2, 3, 4, 5 and 6, respectively. Endogenous S sequences are labeled above the construct bands. Arrows highlight recombinant vector bands that co-hybridize with $S\mu$ and $S\gamma_{2b}$ probes.

by a mechanism other than complete deletion [43]. The absence of retrovector S segment recombination in the A39γ1.1tk⁻ hybridoma line suggests that the normal process of S region recombination is not involved in their endogenous C_H gene rearrangements which could conceivably also occur by sister chromatid or homologous chromosome exchange.

Retrovector-Mediated Switch Recombination Requires both Sµ and Sγ₂ᵦ Sequences

To determine whether the Sµ segment alone might be sufficient for initiating switch recombination, we prepared the ZN(Sµ)tk retroviral construct which lacks an Sγ₂ᵦ segment. The Sµ region could conceivably act as a nucleation site for progressive 5′ to 3′ deletions resulting in *Htk* deletion. Alternatively, the Sµ segment along with S-like sequences fortuitously located in *Htk* coding or 5′-flanking *neo* sequences could facilitate the deletion of the intervening *Htk* gene. ZN(Sµ)tk virus was prepared and used to infect 18-8tk⁻, 38B9tk⁻ and NIH3T3tk⁻ cells. Fluctuation frequencies for the appearance of G-418/BrdU-resistant clones are shown in table 2. The rates of HTK phenotype loss in each of the cell lines were $1-2 \times 10^4$-fold lower than observed for ZN(Sµ/γ₂ᵦ)tk1-infected 300-18tk⁻ cells but essentially the same as seen for Ltk⁻ cells infected with the latter retrovector (see table 1). Genomic DNAs from clones of G-418/BrdU-resistant, ZN(Sµ)tk-infected 18-8tk⁻ and 38B9tk⁻ cells were restricted with BamHI and submitted to Southern hybridization with an HTK probe. The ZN(Sµ)tk1 proviral *Htk* gene was present in an unrearranged context in each of the clones and the vector's Sµ segment was also retained in its original context [29] (data not shown).

Table 2. Loss of HTK phenotype in cells harboring ZN(Sµ)tk, ZN(Sµ/c-*myc*)tk and ZN(Sµ/Sγ₂ᵦ)tk2 proviruses

Cell lines	Loss of HTK phenotype/cell generation	Loss of HTK phenotype relative to ZN(Sµ/Sγ₂ᵦ)tk1-infected 300-18 cells
ZN(Sµ)tk		
18-8tk⁻	5×10^{-8}	1.2×10^{-4}
38B9tk⁻	1×10^{-7}	2.5×10^{-4}
NIH3T3tk⁻	5×10^{-8}	1.2×10^{-4}
ZN(Sµ/c-*myc*)tk		
38B9tk⁻	3×10^{-7}	7.5×10^{-4}
ZNS(Sµ/Sγ₂ᵦ)tk2		
18-8tk⁻	7×10^{-6}	1.7×10^{-2}
38B9tk⁻	2×10^{-6}	5.0×10^{-3}
NIH3T3tk⁻	3×10^{-8}	7.5×10^{-5}

The murine c-*myc* gene on chromosome 15 was found to recombine with the immunoglobin C_H locus on chromosome 12 in plasmacytomas [37, 44]; these aberrant rearrangements usually involved recombinations with the C_H S regions [45, 46], implying that switch recombinase activities may be responsible for the breakage of the c-*myc* locus. These c-*myc* sequences are not particularly related to S segments [47], though some S-like sequences have been noted in this portion of the c-*myc* gene [48]. We prepared the ZN(Sμ/c-*myc*)tk vector to determine if switch recombinase activity could recombine Sμ with 5′-flanking and first exon sequences of the murine c-*myc* gene. The ZN(Sμ/c-*myc*)tk vector contains the same sequences as ZN(Sμ)tk in addition to a 1.7-kb BglII c-*myc* fragment (which is most often involved in chromosome translocations with the C_H S regions) [45] in place of the $Sγ_{2b}$ segment of ZN(Sμ/Sγ$_{2b}$)tk1. 38B9tk$^-$ cells were infected with ZN(Sμ/c-*myc*)tk virus and independent populations of cells were grown in G-418/BrdU selective media and submitted to fluctuation frequency analysis. As shown in table 2, their rate of HTK phenotype loss was similar to that of cells harboring the retrovector with only the Sμ segment. Southern analyses revealed no evidence of Sμ/*myc* recombination [29]. The c-*myc* gene would appear to be no more prone to switch recombination than any other random DNA segment. Furthermore, these findings indicate that the retrovector's Sμ and $Sγ_{2b}$ sequences are both essential for mediating *Htk* gene deletions.

Switch Recombination with a Retrovector Containing Sμ and an Inverted $Sγ_{2b}$ Segment

The ZN(Sμ/$Sγ_{2b}$)tk2 virus contains the same sequences as ZN(Sμ/$Sγ_{2b}$)tk1 except that the $Sγ_{2b}$ sequences of the tk2 vector are in the opposite orientation. This construct would permit the detection of recombinations between S segments of opposite polarity. Fluctuation frequency analysis revealed that 18-8tk$^-$ cells harboring this inversion switch-substrate vector lost their HTK phenotype with a rate 170-fold below that of the 300-18tk$^-$-line containing the original deletion substrate retrovector. On the other hand, NIH3T3tk$^-$ cells harboring the tk2 vector lost their HTK phenotype at a very low rate which was comparable to that observed with the tk1 vector (see table 2). Southern blots of DNAs harvested from G-418/BrdU-resistant, ZN(Sμ/$Sγ_{2b}$)tk2-infected 18-8tk$^-$ and 38B9tk$^-$ clones revealed *Htk* gene deletions and rearranged, Sμ/$Sγ_{2b}$ co-hybridizing bands (see fig. 5 for 18-8tk$^-$ data). As shown in figure 6, the sizes of the recombinants, with one exception, were within the predicted size range of 0.3–1.8 kb for recombinations mediated by the S segment repetitive sequences.

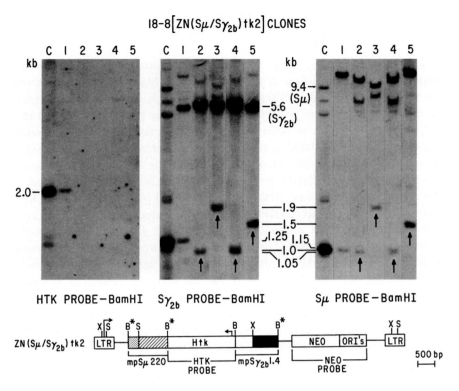

Fig. 5. Southern blot of BamHI-restricted G-418/BrdU-resistant clones of 18-8tk⁻
cells harboring the ZN(Sμ/Sγ₂ᵦ)tk2 retrovector sequentially hybridized to HTK, Sγ₂ᵦ and
Sμ probes. Lane C = BamHI-cut ZN(Sμ/Sγ₂ᵦ)tk2 vector DNA; lane 1 = G-418-resistant
population of infected cells prior to BrdU selection; lanes 2–5 = 4 independent G-418/
BrdU-selected clones of infected 18-8tk⁻ cells. Bands corresponding to endogenous
S sequences are indicated and a diagram of the retrovectors is provided below the blot.
B = BamHI; B* = BamHI by linker modification; S = SstI; X = XbaI.

Sequences of ZN(Sμ/Sγ₂ᵦ)tk1 Recombination Sites in Clones of G-418/BrdU-Resistant 18-8tk⁻ Cells

To define the precise molecular details of the S-region-mediated ZN(Sμ/
Sγ₂ᵦ)tk1 recombinations in pre-B cells, genomic DNAs of 6 independent
clones of infected 18-8tk⁻ cells were restricted with SstI and the recombinant
proviruses recovered by plasmid rescue or by COS cell fusion [29]. Plasmids
recovered from each of these lines (NB29, NB1, NB27, NB32, NB7 and NB3)
were restriction-mapped to confirm that these structures reflected their

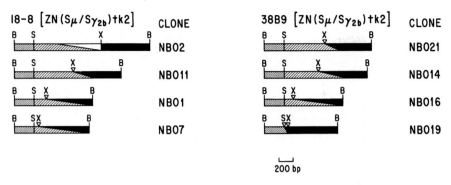

Fig. 6. Southern blot maps of 8 ZN(Sμ/Sγ$_{2b}$)tk2 recombinants derived from infected clones of G-418-BrdU-resistant clones. Open triangles signify that the indicated restriction site was deleted. B = BamHI; X = XbaI; S = SstI.

original genomic contexts prior to rescue and cloning in bacteria. Sequences in the vicinity of the recombination sites and their locations in the parental vector's Sμ and Sγ$_{2b}$ segments are presented in figure 7. The recombinations did not involve S region YAGGTTG sequence motifs except for clone NB7 whose recombination site resides within a 5 out of 7 YAGGTTG match positioned just 3' of the SstI site within the Sμ segment. The NB27 clone recombined in the 3'-terminal 49mer repeat of the Sγ$_{2b}$ segment. The NB29 and NB1 recombination sites were located in a portion of the Sγ$_{2b}$ region that consists of larger, higher-order repeats postulated to function in switch recombination [49]. The NB32 recombinant possessed a complicated structure consisting of 3 independent recombination events. The Sμ-Sγ$_{2b}$ switch in NB32 took place between the repetitive portions of the Sμ and Sγ$_{2b}$ segments. Another independent recombination also occurred within the Sμ segment which did not encompass Sγ$_{2b}$ sequences. Similarly, a third independent deletion was found within the Sγ$_{2b}$ segment. An additional 22-basepair (bp) of Sγ$_{2b}$-related sequences was also present at this latter recombination site which could not be aligned with the Sμ and Sγ$_{2b}$ segments of the original vector. Similar types of sequence additions had been noted at an S region recombination site in a genomic clone isolated from an LPS B cell blast population [50] and, more recently, nearby some of the Sμ/Sα recombinant joints isolated from switched subclones of the I-29 murine B lymphoma line [51]. The events causing the appearance of these extra nucleotides are not clear though some of the structure may be accounted for by multiple small deletions or possibly by DNA repair processes.

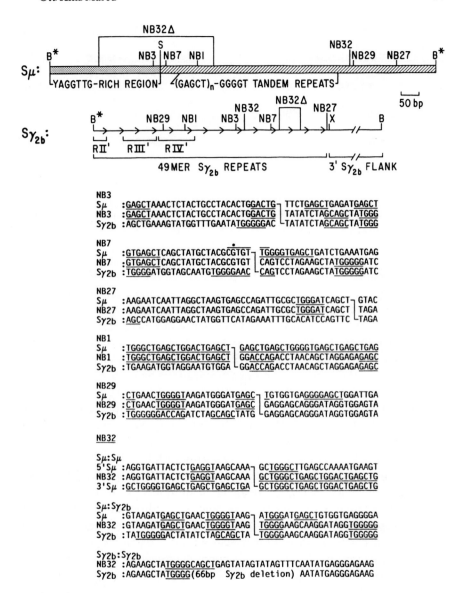

Fig. 7. Sequence analysis of 6 recombined ZN(Sμ/Sγ$_{2b}$)tk1 proviruses. Maps of the Sμ and Sγ$_{2b}$ segments in the retroviral vector and the locations of the recombination sites in each of the clones are presented above the sequences. The Sμ map shows the YAGGTTG-rich region of Sμ as a stippled bar whereas the (GAGCT)$_n$GGGGG region is denoted by the striped bar. The tandem repeat sequences within Sμ are bracketed within

Although there were no consistent consensus sequences found near the sites of switch recombination in each of these 6 independent clones, there were characteristic S region sequence motifs clustered nearby most of the recombining sequences. These GAGCT, GGGGT, TGGGG, ACCAG and GCAGC common repeats are indicated in figure 7. Even though such S-like sequences are near these recombination sites, the DNA sequences flanking the breakpoints in each strand bear no consistent homology. Although the sites of recombination contain several motifs characteristic of the Sμ repeat structure, they lack the solid block of tandem repeats that comprise the bulk of the Sμ region [7]. Furthermore, the Sμ/Sγ$_{2b}$ recombination sites in Sμ reside in the most 5′ or 3′ portions of the Sμ segment which constitute the least-repetitive part of the Sμ region. In addition, there are no preferential recombination sites either within the entire repetitive Sγ$_{2b}$ region or within individual 49mer Sγ$_{2b}$ repeats. Because we do not find S segment tandem repeat sequences at the precise sites of switch recombination, our findings would support the view that these sequences play an essential though indirect role in C$_H$ gene switching.

Status of the Endogenous Sμ and Sγ$_{2b}$ Sequences in Pre-B Lines Undergoing Retrovector S Segment Recombinations

The endogenous Sμ and Sγ$_{2b}$ sequences in the various cell lines were examined to determine if they coordinately rearranged along with the retrovectors. The 18-8tk⁻ cell line was shown to switch its isotype in culture and also to recombine its endogenous Sμ sequences as a prelude to the switch event, in a similar manner to LPS-treated spleen cells [4, 11, 14]. 18-8tk⁻ clones which rearranged the ZN(Sμ/Sγ$_{2b}$)tk1 and tk2 vectors generally contained rearranged Sμ sequences. In 12 of 14 tk1 clones and 3 of 4 tk2 clones, endogenous Sμ rearrangements are present on either one or both alleles. However, similar clones of 300-18tk⁻ and 38B9tk⁻ cells did not show such

---

the striped region. The Sγ$_{2b}$ map shows the tandem 49mer repeats as tandem arrows and also the locations of longer repeat structures (RII′, RIII′ and RIV′) [49]. Clone NB32 contained two deletions, denoted by the connected lines with open triangles, in addition to an Sμ/Sγ$_{2b}$ recombination. Sequences are presented as the germ line Sμ and Sγ$_{2b}$ sequences [8] with the recombinant sequences in between. Each switch recombination is marked by a line connecting the two germ line sequences. Sequence motifs common to S regions are underlined while a 5 out of 7 match to a YAGGTTG consensus sequence is marked by an overhead bar with a dot. B = BamHI; B* = BamHI by linker modification; S = SstI; X = XbaI.

endogenous Sμ recombinations. Furthermore, no evidence for endogenous Sγ$_{2b}$ rearrangements was apparent in any of these cell lines.

Identification of Ubiquitous and B-Cell-Specific Sμ-Binding Proteins

In another approach to define the molecular requirements for switch recombination, we have recently employed a gel electrophoretic DNA-

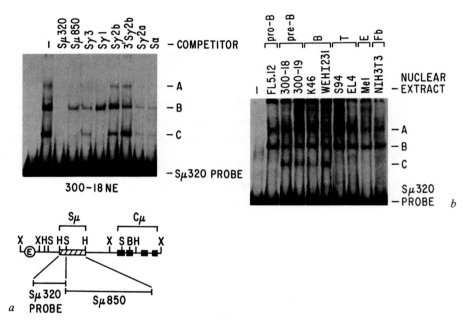

Fig. 8. a Formation of DNA-protein complexes between factors in a crude nuclear extract of 300-18 cells to a 320-bp Sμ probe are differentially competed by other S segments. Binding reactions contained 1 ng of a 320-bp fragment of the Sμ region and 5 μg of 300-18 cell crude nuclear extract. The 320-bp Sμ probe is located 5′ of the Sμ tandem repeat region (Sμ850) and the other S segment competitor DNAs, which are all derived from the repetitive portions of the indicated S regions (with the exception of 3′-Sγ$_{2b}$ which reresides between Sγ$_{2b}$ and Cγ$_{2b}$, were present in a 200-fold molar excess with respect to the Sμ320 probe. The sizes of the Sγ$_3$, Sγ$_1$, Sγ$_{2b}$, Sγ$_{2a}$ and Sα DNAs were 0.8, 2.1, 0.6, 2.6 and 2.5 kb, respectively [9]. The positions of migration of 3 major DNA-protein complexes are indicated by A, B and C. *b* Gel retardation assays with the Sμ320 probe and crude nuclear extracts prepared from various cell lines representative of different stages of B cell differentiation, mature T, erythroid and fibroblasts. Binding reactions were performed with 1 ng of the Sμ320 probe and 5 μg of each crude nuclear extract. A, B and C represent the positions of migration of DNA-protein complexes observed with the 300-18 extract in *a*.

protein-binding assay [38] to detect nuclear factors which specifically bind to S segments. Some of our preliminary findings are presented in figure 8a and b where we have screened crude nuclear extracts prepared from pro-B, pre-B, mature B, mature T, erythroid and fibroblast cells for Sμ-specific binding factors (fig, 8b). Three major DNA-protein complexes (designated A, B and C) are observed with a 320-bp DNA fragment derived from the 5' portion of the Sμ region. The nuclear factors responsible for the A and B complexes are expressed in all of the cell lines examined while the C complex protein(s) are only expressed in the B cell lines. Similar gel shift assays indicate that the C complex is not detected in nuclear extracts prepared from plasma cell tumor lines, suggesting that some of these factors may only be expressed in early- and intermediate-stage B cells (data not shown). Different S segments were added to the reactions in a 200-fold molar excess and each competed for Sμ320 probe binding to different degrees (fig. 8a). The formation of all 3 complexes was abolished in the presence of an excess of homologous competitor. The B-cell-specific C complex was effectively competed by the repetitive portion of the Sμ region (Sμ850) and by varying degrees with $S\gamma_3$, $S\gamma_1$, $S\gamma_{2b}$, $S\gamma_{2a}$ and Sα sequences. Each of these S segment competitor DNAs contained from 0.6 to 2.6 kb of S repeat sequences. A DNA fragment located in between the repetitive portion of $S\gamma_{2b}$ and the $C\gamma_{2b}$ gene (3-$S\gamma_{2b}$) was not an effective competitor. These results suggest that the Sμ binding sites for the factor(s) involved in complex C may also be present in the repetitive portions of other S segments. DNA footprinting and competition assays performed with synthetic oligonucleotides indicate that sequences found in the repetitive portion of the Sμ region [(GAGCT)$_n$GGGGT] constitute a portion of the binding site required for the B-cell-specific C complex but other nonrepetitive Sμ sequences are also essential for complex stability (data not shown). These nuclear factors may either facilitate appropriate S segment alignments (synapsis) as a prelude to recombination or conceivably play a more direct role in the rearrangement process.

Discussion

Switch Recombinase Activity in A-MuLV-Transformed Pre-B Cell Lines

Switch recombination induced the deletion of the *Htk* gene in ZN(Sμ/$S\gamma_{2b}$)tk1 proviruses in 3 A-MuLV pre-B lines (18-81, 38B9 and 300-18) [28, 29] but was not observed in NIH3T3tk⁻ and Ltk⁻ fibroblasts [28] or in the A39γ1.1tk⁻ hybridoma [29]. The documented rate for spontaneous gene mutations in hamster cells, 4×10^{-8} events/cell/generation [52], is consistent with the rates of HTK phenotype loss observed here for Ltk⁻ and A39Rγ1.1tk⁻ cells. Southern blot analysis of ZN(Sμ/Sγ$_{2b}$)tk1-infected A39Rγ1.1tk⁻ cells did not reveal switch recombinase activity. The rates of HTK phenotype loss in the 38B9 and 18-8 lines ranged from 1 to 5×10^{-5} events/cell/generation and required BrdU selection to identify clones of cells with proviral *Htk* gene deletions [28]. However, the 300-18tk⁻ pre-B line inactivated the proviral *Htk* gene by Sμ-region-mediated deletion at a minimum estimated frequency of 1 in 2,500 cells per generation. Consistent with their high frequency of HTK phenotype loss, heterogeneous populations of ZN(Sμ/Sγ$_{2b}$)tk1-infected, G-418-resistant 300-18tk⁻ cells also demonstrated a high recombinational activity in the absence of BrdU selection. Some populations showed many rearranged bands while others displayed a more modest level of recombination. The rearranged bands co-hybridized with both Sμ and Sγ$_{2b}$ probes, demonstrating that they were Sμ/Sγ$_{2b}$ fusions. The fluctuations in the amount of rearrangement activity in the different populations are predicted for an ongoing random event by the Luria-Delbruck fluctuation test [34]. Events occurring early in the expansion of the populations are, therefore, magnified with respect to events occurring later on during population expansion. The higher rate of recombination in 300-18tk⁻ compared to 18-8tk⁻ cells may be due to a higher level of switch recombinase activity in 300-18tk⁻ or possibly to a lack of viability of 18-8tk⁻ clones which have also undergone an endogenous Cμ to Cγ$_{2b}$ switch [13, 27].

Sequences Required for Switch Recombination

Switch deletions in B cells require two distinct S segments. A single Sμ segment does not serve as a possible target for this B-cell-specific recombinase activity and the 5'-flanking and intragenic sequences of the murine c-*myc* gene are also ineffective in switch recombination with the S region. Switch recombination was also observed between Sμ and Sγ$_{2b}$ segments of opposite polarity albeit at somewhat lower frequencies. It remains to be determined whether the latter rearrangements result from simple deletions between the inverted S segments or are a two-step process with an inversion preceding a deletion event.

Sequence analysis of 5 independent ZN(Sμ/Sγ$_{2b}$)tk proviral recombinants derived from 3 pre-B cell lines consistently revealed S sequence repeats near the sites of rearrangement; but the Sμ and Sγ$_{2b}$ sequences were not homologous to each other at the sites of recombination. In addition, these recombinations took place within Sμ sequences which flank most of the tandemly repetitious portion of the Sμ region. Therefore, the tandemly repeated $(GAGCT)_n GGGGT$ core sequences of the Sμ region remained intact, suggesting that this region may be protected from recombination. The Sμ region YAGGTTG consensus motif was not found near these recombination sites, but we did observe 1 case of a recombination site within a 5 out of 7 match to a YAGGTTG consensus. The significance of the latter YAGGTTG-associated recombination is unknown though it would appear to be an infrequent occurrence. The lack of homology or a consistent consensus sequence positioned at the precise sites of recombination suggests a mechanism in which multiple copies of switch sequences (i.e. GAGCT-, GGGGT- and YAGGTG-like motifs) are indirectly required for switch recombination by facilitating the alignment of the recombining sequences. A ubiquitous eukaryotic protein with a recombinase-A-like activity has been reported [53] which also has the ability to synapse two partially homologous pieces of DNA [Fred Mills, pers. commun.]. A B-cell-specific S-segment-binding factor with similar properties could conceivably mediate the C_H class switch by facilitating the synapsis of two accessible S regions, but the subsequent recombination event would take place by a nonhomologous mechanism [54, 55]. This model and our results concur with the locations of C_H gene switch sites in plasmacytomas which usually reside 5′ of the repetitious portion of the S region [1, 6].

Does Switch Recombination Occur in a Direct or a Stepwise Manner?

The first evidence for a stepwise or possibly successive switch-deletion mechanism came from the early identification of chimeric switch recombination sites in plasmacytomas and hybridomas [41, 56]. A recent detailed molecular analysis of the Cμ to Cα switch recombination sites in the I-29 murine B lymphoma line has also detected multiple sequence deletions and duplications near the sites of recombination, which may have occurred during or subsequent to the actual switch event [51]. Furthermore, the 18-8tk⁻ cell line is known to delete a portion of its endogenous Sμ repeats prior to switching [11, 27] and LPS-stimulated B cells also demonstrate intra-S deletions prior to the switch event [4, 50, 57, 58].

The sequence of the 18-8tk⁻ NB32 clone's recombinant S segments revealed an interesting structure indicating the occurrence of multiple recombination events. The Sμ/Sγ$_{2b}$ recombination which deleted the *Htk* gene was flanked by intra-S region recombinations in both the Sμ and Sγ$_{2b}$ sequences, and therefore provides direct evidence that S region recombinations could involve more than one deletion event. Inter-S segment recombinations may be preceded by intra-S region deletions. The formation of inter-S region synapses could be rate-limiting events which only occur after several intra-Sμ region deletions. Evidence for the majority of such intra-S region deletions would be lost in the majority of the previously examined C_H switches in plasmacytomas, since the use of the 5'-Sμ region in the bulk of these events would result in the deletion of the initial intra-Sμ rearrangement residing 3' of the final recombination site. However, in cases where the recombinations took place within the 3'-Sμ region (as shown here for the NB32 18-8tk$^-$ clone), intra-Sμ recombinations were also found 5' of the Sμ site [14, 50].

Is the Presence of Switch Recombinase Activity Sufficient for C_H Gene Recombination?

The 38B9tk$^-$ cell line was previously shown to undergo S-region-mediated recombination within the ZN(Sμ/Sγ_{2b})tk1 provirus without endogenous Sμ recombination [28]; even though the S sequences of the ZN(Sμ/Sγ_{2b})tk1 retrovirus recombined with the highest frequency in the 300-18tk$^-$ pre-B line, its corresponding endogenous C_H genes appear to be recombinationally inert. A reasonable assumption would be that endogenous S segments recombine at a much lower frequency than those introduced on a retrovector. The lack of endogenous intra-Sμ recombination in the 38B9 and 300-18 lines is probably not due to a lack of S segment accessibility since these cell lines express either sterile or functional Cμ transcripts, respectively. The 38B9 cell line does not express Cμ protein but Northern blot analyses have revealed the presence of sterile transcripts [59, 60]. The infected 300-18tk$^-$ cell lines did not reveal endogenous Sμ rearrangements when a BamHI Southern blot was hybridized with either Sμ or other Cμ intron probes residing 5' of the Sμ region (data not shown). Additionally, the 300-18tk$^-$ G-418/BrdU-resistant clones did not stain positively with TRIC-conjugated anti-γ_{2b} antibody nor did the clones stain positively with FITC-conjugated anti-κ. However, the 300-18tk$^-$ cell line stained positively with FITC-conjugated anti-μ and, therefore, transcribes the Sμ sequences; others have shown that this line also expresses sterile Cγ_{2b} transcripts [61, 62].

The absence of endogenous Sμ or Sμ to Sγ_{2b} rearrangements in the 38B9 and 300-18 lines suggests that their endogenous C_H alleles are subject to some type of control to which the S region retrovector is refractory, while the 18-8tk$^-$ cell line shows a high correlation of endogenous Sμ and ZN(Sμ/Sγ_{2b})tk1 vector rearrangement [28]. Even though the 18-8 Sμ region is recombinationally active, endogenous Sμ to Sγ_{2b} switches in 18-8 probably occur at a much lower frequency than such intra-Sμ deletions and have also been hypothesized to require $S\gamma_{2b}$ accessibility [26, 61, 62]. A similar phenomenon has been suggested to explain infrequent $C\mu$ to $C\gamma_{2b}$ switches in the 300-18 line [26, 61, 62]. Therefore, the frequency of endogenous S region rearrangement does not provide a good indication of the level of switch recombinase activity. The $S\mu/S\gamma_{2b}$ retroviral vector may be a better substrate for the switch recombinase in these cells due to the high level of transcriptional activity provided by the retroviral LTR and the close proximity of the retrovector's S sequences. Alternatively, one or more other factors may be required for endogenous S region recombination in addition to the presence of switch recombinase and accessible S segment target sequences. It will be important to determine the role(s) of ubiquitous and B-cell-stage-specific S-region-binding proteins in this phenomenon.

References

1 Marcu KB: Immunoglobulin heavy-chain constant region genes. Cell 1982;29:719–721.
2 Honjo T: Immunoglobulin genes. Annu Rev Immunol 1983;1:499–528.
3 Shimizu A, Honjo T: Immunoglobulin class switching. Cell 1984;36:801–803.
4 Radbruch A, Burger C, Klein S, et al: Control of immunoglobulin class switch recombination. Immunol Rev 1986;89:69–83.
5 Gritzmacher CA: Molecular aspects of heavy-chain class switching. Crit Rev Immunol 1989;9:173–200.
6 Marcu KB. Lang RB, Stanton LW, et al: A model for the molecular requirements of immunoglobulin heavy-chain class switching. Nature 1982;298:87–89.
7 Nikaido T, Nakai S, Honjo T: Switch region of immunoglobulin $C\mu$ gene is composed of simple tandem-repetitive sequences. Nature 1981;292:845–848.
8 Kataoka T, Miyata T, Honjo T: Repetitive sequences in class switch recombination regions of immunoglobulin heavy-chain genes. Cell 1981;23:357–368.
9 Stanton LW, Marcu KB: Nucleotide sequences and properties of the murine γ_3 immunoglobulin heavy-chain gene switch region: Implications for successive $C\gamma$ gene switching. Nucleic Acids Res 1982;10:5993–6006.
10 Davis MM, Kim SK, Hood LE: DNA sequences mediating class switching in alpha immunoglobulins. Science 1980;209:1360–1365.
11 Alt FW, Rosenberg N, Casanova R, et al: Immunoglobulin heavy chain expression and class-switching in a murine leukemia cell line. Nature 1982;296:325–331.
12 Akira S, Sugiyama H, Yoshida N, et al: Isotype switching in murine pre-B cell lines. Cell 1983;34:545–556.
13 Burrows PD, Beck-Engeser GB, Wabl MR: Immunoglobulin heavy chain class switching in a pre-B cell line is accompanied by DNA rearrangement. Nature 1983;306:243–246.
14 DePinho R, Kruger K, Andrews N, et al: Molecular basis of heavy-chain class

switching and switch region deletion in an abelson virus-transformed cell line. Mol Cell Biol 1984;4:2905–2910.

15 Yancopoulos GD, DePhino RA, Zimmerman KA, et al: Secondary genomic rearrangement events in pre-B cells: $V_H DJ_H$ replacement by a LINE-1 sequence and directed class switching. EMBO J 1986;5:3259–3266.

16 Stavnezer J, Marcu KB, Sirlin S, et al: Rearrangements and deletions of immunoglobulin heavy chain genes in the double-producing B cell lymphoma I.29. Mol Cell Biol 1982;2:1002–1013.

17 Stavnezer J, Sirlin S, Abbott J: Induction of immunoglobulin isotype switching in cultured I-29 B lymphoma cells: Characterization of the accompanying rearrangements of heavy-chain genes. J Exp Med 1985;161:577–601.

18 Eckhardt LA, Tilley SA, Lang RB, et al: DNA rearrangements in MPC-11 immunoglobulin heavy chain class-switch variants. Proc Natl Acad Sci USA 1982;79:3006–3010.

19 Eckhardt LA, Birshtein BK: Independent immunoglobulin class switch events occurring in a single myeloma cell line. Mol Cell Biol 1985;5:856–869.

20 Radbruch A, Sablitzky F: Deletion of Cμ genes in mouse B lymphocytes upon stimulation with LPS. EMBO J 1983;2:1929–1935.

21 Stavnezer J, Abbott J, Sirlin S: Immunoglobulin heavy-chain switching in cultured I.29 murine B lymphoma cells: Commitment to an IgA or IgE switch. Curr Top Microbiol Immunol 1984;113:109–116.

22 Stavnezer J, Sirlin S: Specificity of immunoglobulin heavy-chain switch correlates with activity of germline heavy chains prior to switching. EMBO J 1986;5:95–102.

23 Alt FW, Blackwell TK, Yancopoulos G: Development of the primary antibody repertoire. Science 1987;238:1079–1087.

24 Perry RP, Kelley DE, Coleclough C, et al: Organization and expression of immunoglobulin genes in fetal liver hybridomas. Proc Natl Acad Sci USA 1981;78:247–251.

25 Yancopoulos GD, Alt FW: Developmentally controlled and tissue-specific expression of unrearranged V_H gene segments. Cell 1985;40:271–281.

26 Yancopoulos GD, Blackwell TK, Suh H, et al: Introduced T cell receptor variable region gene segments recombine in pre-B cells: Evidence that B and T cells use a common recombinase. Cell 1986;44:251–259.

27 Blackwell TK, Moore MW, Yancopoulos GD, et al: Recombination between immunoglobulin variable region gene segments is enhanced by transcription. Nature 1986;324:585–589.

28 Ott DE, Alt FW, Marcu KB: Immunoglobulin heavy-chain switch region recombination within a retroviral vector in murine pre-B lymphoid cells. EMBO J 1987;6:577–584.

29 Ott DE, Marcu KB: Molecular requirements for immunoglobulin heavy-chain constant region gene switch-recombination revealed with switch-substrate retroviruses. Int Immunol 1989;1:582–591.

30 Alt FW, Yancopoulos GD, Blackwell TK, et al: Ordered rearrangement of immunoglobulin genes. EMBO J 1984;3:1209–1219.

31 Mann R, Mulligan RC, Baltimore D: Construction of a retrovirus packaging mutant and its use to produce helper-free defective retrovirus. Cell 1983;33:153–159.

32 Cepko CL, Roberts BE, Mulligan RC: Construction and applications of a highly transmissable murine retrovirus shuttle vector. Cell 1984;37:1053–1062.

33 Miller AD, Law MF, Verma IM: Generation of helper-free amphotropic retroviruses

that transduce a dominant-acting, methotrexate-resistant dihydrofolate reductase gene. Mol Cell Biol 1985;5:431–437.

34 Luria SE, Delbruck M: Mutations of bacteria from virus sensitivity to virus resistance. Genetics 1943;28:491–511.

35 Rigby PWJ, Dieckmann M, Rhodes C, et al: Labeling deoxyribonucleic acid to high specific activity in vitro by nick-translation with DNA polymerase I. J Mol Biol 1977; 113:237–251.

36 Hu NT, Messing J: The making of strand-specific M13 probes. Gene 1982;17:271–277.

37 Marcu KB, Harris LJ, Stanton LW, et al: Transcriptionally active c-*myc* oncogene is contained within NIARD, a DNA sequence associated with chromosome transloca-tions in B cell neoplasia. Proc Natl Acad Sci USA 1983;80:519–523.

38 Singh H, Sen R, Baltimore D, et al: A nuclear factor that binds to a conserved sequence motif in transcriptional control elements of immunoglobulin genes. Nature 1986;319:154–158.

39 Dignam JD, Lebowitz RM, Roeder RG: Accurate transcription initiation by RNA polymerase II in a soluble extract from isolated mammalian nuclei. Nucleic Acids Res 1983;11:1475–1489.

40 Varmus HE, Swanstrom R: Replication of retroviruses; in Weiss R, Teich N, Varmus H, Coffin J (eds): RNA Tumor Viruses. Cold Spring Harbor, Cold Spring Harbor Laboratory, 1984, pp 75–134.

41 Nikado T, Yamawaki-Kataoka Y, Honjo T: Nucleotide sequences of switch regions of immunoglobulin Cγ and Cε genes and their comparison. J Biol Chem 1982;257: 7322–7329.

42 Blackwell TK, Alt FW: Site-specific recombination between immunoglobulin D and J_H segments that were introduced into the genome of a murine pre-B cell line. Cell 1984;37:105–112.

43 Stringer JR, Kuhn RM, Newman JL, et al: Unequal homologous recombination between tandemly arranged sequences stably incorporated into cultured rat cells. Mol Cell Biol 1985;5:2613–2622.

44 Harris LJ, D'Eustachio P, Ruddle FH, et al: DNA sequence associated with chromosome translocations in mouse plasmacytomas. Proc Natl Acad Sci USA 1982; 79:6622–6626.

45 Cory S: Activation of cellular oncogenes in hematopoietic cells by chromosome translocation. Adv Cancer Res 1986;47:189–234.

46 Marcu KB: Regulation of expression of the c-*myc* proto-oncogene. Bioassays 1987; 6:28–32.

47 Bernard O, Cory S, Gerondakis S, et al: Sequence of the murine and human cellular *myc* oncogenes and two modes of *myc* transcription resulting from chromosome translocation in Burkitt lymphoid tumors. EMBO J 1983;2:2375–2383.

48 Dunnick W, Shell BE, Dery C: DNA sequence near the site of reciprocal recombina-tion between a c-*myc* oncogene and an immunoglobulin switch region. Proc Natl Acad Sci USA 1983;80:7269–7273.

49 Wu TT, Reid-Miller M, Perry HM, et al: Long identical repeats in the mouse gamma-2b switch region and their implications for the mechanism of class-switching. EMBO J 1984;3:2033–2040.

50 Winter E, Krawinkle U, Radbruch A: Directed Ig class switch recombination in activated murine B cells. EMBO J 1987;6:1663–1671.

51   Dunnick W, Wilson M, Stavnezer J: Mutations, duplication and deletion of recombined switch regions suggest a role for DNA replication in the immunoglobulin heavy-chain switch. Mol Cell Biol 1989;9:1850–1856.

52   Nalbantoglu J, Phear G, Meuth M: DNA sequence analysis of spontaneous mutations at the APRT locus of hamster cells. Mol Cell Biol 1987;7:1445–1449.

53   Hsieh P, Meyn SM, Camerini-Otero RD: Partial purification and characterization of a recombinase from human cells. Cell 1986;44:885–894.

54   Anderson RA, Kato S, Camerini-Otero RD: A pattern of partially homologous recombination in mouse L cells. Proc Natl Acad Sci USA 1984;81:206–210.

55   Roth DB, Wilson JH: Non-homologous recombination in mammalian cells: Role for short sequence homologues in joining reaction. Mol Cell Biol 1986;6:4295–4304.

56   Obata M, Kataoka T, Nakai S, et al: Structure of a rearranged γ_1-chain gene and its implication to immunoglobulin class-switch mechanism. Proc Natl Acad Sci USA 1981;78:2437–2441.

57   Hurwitz J, Coleclough C, Cebra J: C_H gene rearrangements in IgM-bearing B cells and in the normal splenic DNA component of hybridomas making different isotypes of antibody. Cell 1980;22:349–359.

58   Hurwitz J, Cebra JJ: Rearrangements between the immunoglobulin heavy-chain gene J_H and $C\mu$ regions accompany normal B lymphocyte differentiation in vitro. Nature 1982;299:742–744.

59   Lennon GG, Perry RP: $C\mu$-containing transcripts initiate heterogeneously within the Igh enhancer region and contain a novel 5′-nontranslatable exon. Nature 1985; 318:475–477.

60   Reth MG, Alt FW: Novel immunoglobulin heavy-chains are produced from DJ_H gene segment rearrangements in lymphoid cells. Nature 1984;312:418–423.

61   Lutzker S, Alt FW: Structure and expression of germ-line immunoglobulin γ_{2b} transcripts. Mol Cell Biol 1988;8:1849–1852.

62   Lutzker S, Rothman P, Pollock R, et al: Mitogen- and IL-4-regulated expression of germ-line $Ig\gamma_{2b}$ transcripts: Evidence for directed heavy-chain class switching. Cell 1988;53:177–184.

Prof. Kenneth B. Marcu, PhD, Department of Biochemistry and Cell Biology, SUNY at Stony Brook, Stony Brook, NY 11794–5215 (USA)

Sorg C (ed): Molecular Biology of B Cell Developments.
Cytokines. Basel, Karger, 1990, vol 3, pp 85–108

Regulation of Immunoglobulin Expression: Role of Lymphoid-Specific and Ubiquitous Nuclear Factors

Patrick Matthias[1]

Whitehead Institute for Biomedical Research,
Nine Cambridge Center, Cambridge, Mass., USA

The antibody molecules, or immunoglobulins, are made exclusively by a category of dedicated cells, the B lymphocytes. During B lymphocyte development, immunoglobulin expression changes qualitatively and quantitatively, both at the level of RNA and protein. Here, I will briefly outline different levels of regulation of immunoglobulin expression and then focus on the transcriptional regulation of immunoglobulin genes and on the transcription factors involved.

Functional immunoglobulin molecules consist of two heavy-chain and two light-chain polypeptides held together by disulfide bonds [1]. In man or mice, one locus in the genome encodes the immunoglobulin heavy chain (Igh) and two different loci code for the immunoglobulin light chains (Igl; κ- or λ-type). In the germ line configuration, however, the various coding segments are too far apart (several hundreds of kilobases [2]) to encode functional immunoglobulin genes. Thus, the first step of commitment to immunoglobulin expression is a DNA rearrangement restricted to B lymphocytes, whereby various DNA pieces are brought together to form a functional gene. This site-specific DNA recombination takes place in several successive steps which serve as landmarks for the differentiation stage of B lymphocytes (fig. 1).

[1] I thank Roger Clerc for help with the MacIntosh and critical comments, and the Swiss National Science Foundation for support.

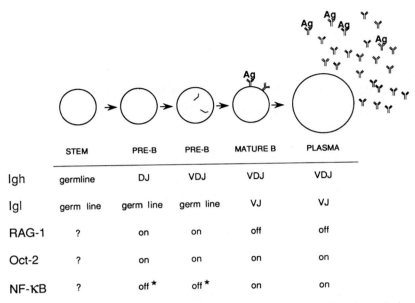

	STEM	PRE-B	PRE-B	MATURE B	PLASMA
Igh	germline	DJ	VDJ	VDJ	VDJ
Igl	germ line	germ line	germ line	VJ	VJ
RAG-1	?	on	on	off	off
Oct-2	?	on	on	on	on
NF-κB	?	off *	off *	on	on

Fig. 1. Simplified representation of B cell differentiation (upper part) and correlation with the DNA structure of the Igh and Igl loci. Germ line indicates that the allele is in the germ line configuration. The mature B cell is represented expressing a functional immunoglobulin at the cell surface. An antigen is symbolized by Ag. The expression of the recombination-activating gene, RAG-1 [6] and of the transcription factors Oct-2 and NF-κB is also indicated in the lower part of the figure. For RAG-1 the data are based on RNA analysis [6]. For Oct-2 and NF-κB the data are mostly derived from binding assays with nuclear extracts and are only qualitative. For instance, Oct-2 is expressed at higher levels in mature B or in plasma cells than in pre-B cells [59], but can be induced in at least 2 pre-B cell lines (70Z-3 and HAFTL-1) by LPS treatment [59, 60]. For NF-κB, 'on' indicates the presence of DNA-binding activity in the cell nucleus; 'off*' indicates that the DNA-binding activity is not present in the nucleus, but can be induced by in vivo treatment of cells or by in vitro treatment of extracts. Examples of cell lines representing the different stages of B cell development and used for these studies are: HAFTL-1, 38B9, 70Z-3 (pre-B cells); WeHi, Namalwa, BJA-B (mature B cells); S194 (plasmacytoma cells) [see ref. 6 for description of the cell lines].

The Igh locus rearranges first, by bringing a diversity (D) segment next to a joining (J) segment, and then juxtaposing to this unit one of several hundred variables (V) segments [2]. This creates a VDJ exon which is separated from the constant (C)-region exons by a large intron (fig. 2a). Subsequently, the Igl locus rearranges its DNA similarly and brings one of many V segments (κ-locus) next to a J segment (there is no D element in that

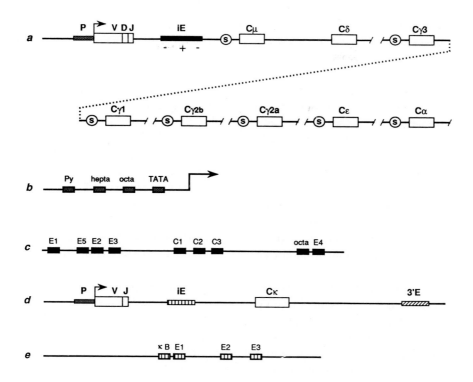

Fig. 2. Structure of the rearranged immunoglobulin loci. *a* Igh locus. The coding regions are depicted by open boxes without indication of exon-intron boundaries. The V, D and J segments are indicated, as well as the regions where switch recombination (S) usually occurs, upstream of the C regions. The solid boxes labeled P and iE represent the immunoglobulin promoter and intron enhancer, respectively. The enhancer downstream of the Cα gene [18] is not indicated. Underneath the intronic enhancer the '–' signs refer to the two regions of the enhancer (positions 1–345 and 566–996) which are implicated in the negative regulation in non-B cells [24–27]. The '+' sign refers to the central region which confers full activity in B cells. The arrow indicates the transcription start site. *b* Enlarged view of the immunoglobulin promoter with the conserved pyrimidine-rich region (Py), the heptamer sequence (hepta: -CTCATGA- [35]), the octamer sequence (octa) and the TATA box indicated. *c* Enlarged view of the Igh enhancer central region (224-bp HinfI fragment) showing the position of the E boxes (E1–E5), the homologies to the SV40 enhancer core sequence (C1–C3), and the octamer site (octa). *d* Igl-κ locus. The stippled box (iE) and the hatched box (3'E) indicate the intronic [19] and 3' enhancers [17], respectively. *e* Structure of the intronic enhancer (465-bp AluI fragment). The position of the NF-κB site is indicated (κB), as well as that of E boxes highly related to those found in the Igh enhancer (E1–E3). Drawings are not to scale.

case [2]; fig. 2d). This combinatorial building up of immunoglobulin genes plays a central role in establishing the enormous diversity of the primary antibody repertoire.

At present, little is known about the molecular mechanism of immunoglobulin recombination, except that it is critically dependent on short conserved sequences flanking the immunoglobulin gene segments (the heptamer and the nonamer sequences [2, 3]) and that recombination correlates with transcription or accessibility of the locus [2, 4]. Furthermore, it has been shown that the same recombination machinery is also carrying out the T-lymphocyte-restricted DNA rearrangement of the T cell receptor genes [2, 5]. Probably the most exciting finding in that field has been the very recent isolation in the laboratory of David Baltimore of a recombination-activating gene, RAG-1, whose expression is sufficient to activate site-specific immunoglobulin recombination of appropriate substrates in non-lymphoid cells [6]. As expected for a recombinase gene, RAG-1 is expressed only in those pre-B cells that actively carry out immunoglobulin rearrangement [6] (fig. 1).

When a B lymphocyte has successfully rearranged both its Igh and Igl loci it can make a functional immunoglobulin molecule which is displayed at the cell surface, where it serves as a 'receptor' for antigen. Such a mature B cell rests until it encounters its cognate antigen. This then triggers proliferation and further differentiation of the B cell which eventually becomes a plasma cell, secreting very large amounts of antibody (fig. 1). The mechanism which regulates the expression of membrane-bound versus secreted immunoglobulin appears to be operating at the level of immunoglobulin transcript polyadenylation [7, 8] or splicing [9].

Furthermore, the steady-state levels of immunoglobulin RNA increase enormously between early B cells and plasma cells [10]. This regulation also seems to be mostly, if not exclusively, at the posttranscriptional level, as the rate of immunoglobulin gene transcription, measured by run-on analysis, remains essentially constant throughout B cell differentiation [11]. Finally, at late stages of an immune response isotype switching and somatic hypermutation can occur. Isotype switching allows the B lymphocyte to make an antibody with the same antigenic specificity (i.e. the same VDJ region) but with a new effector function by using a different C region. This mechanism involves in almost all cases another type of DNA rearrangement which is dependent on repeated sequences found close upstream of the various Igh C regions (fig. 2) [2]. Somatic hypermutation, on the other hand, is a mechanism by which the cell introduces mutations in the V region of the antibody molecule it produces, probably to fine-tune its specificity [12]. How this happens, and what controls the spatial and temporal occurrence of these mutations, is totally unknown at present. Clearly, all these levels of regulation (and the putative feedbacks between them) contribute to the overall B-cell-specific expression of immunoglobulin genes.

Transcriptional Regulation of Immunoglobulin Genes

At least two elements have been identified which control the transcription of immunoglobulin genes: the V region promoter and the enhancer(s) (fig. 2). Each V region has a functional promoter which, however, is only very weakly active in the unrearranged state (i.e. in early pre-B cells [13]). After DNA rearrangement the newly created exon (VDJ or VJ) becomes highly transcribed from the V region promoter. The explanation for that finding is the presence of a transcriptional enhancer in the intron separating the J and C regions of the Igh and Igl-κ locus [14–16] (fig. 2). DNA rearrangement effectively brings the V region promoter under the influence of the already active [11] enhancer. Recently, it has been found that the κ-locus contains a second, stronger enhancer 3' of the C region [17] (fig. 2). The heavy chain locus also contains at least one additional enhancer 25 kb downstream of the Cα gene [18].

The most remarkable finding was that the V region promoter as well as the immunoglobulin enhancers are lymphoid-specific. When linked to a heterologous test gene, these genetic elements direct transcription only in lymphoid (mostly B) cells [14, 15, 17, 21]. A major effort has been devoted to try understanding what makes the immunoglobulin promoter and enhancers active only in B cells. Extensive functional analyses as well as in vivo and in vitro binding studies have identified a large number of DNA motifs and corresponding binding proteins which contribute to the overall activity of the immunoglobulin promoter and enhancers [reviewed in ref. 22 and 23]. Moroever, the analysis of the Igh enhancer has revealed the presence of negative elements flanking the central positively acting region of the enhancer. These negative sequences have been proposed to be reponsible for the complete inactivity of the Igh enhancer in non-lymphoid cells [24–27].

Below I discuss in more detail the contribution of three DNA motifs [the octamer motif, the Ephrussi (E) boxes and the κB site] and their cognate factors to immunoglobulin transcription.

Octamer Motif
The Octamer Paradox. The octamer motif -TNATTTGCAT- or its reverse complement is found in all immunoglobulin promoters (heavy and

light chain) as well as in the Igh enhancer [28, 29]. Moreover, this conserved motif is also associated with 'housekeeping' genes displaying no apparent lymphoid specificity, such as some small nuclear (sn) RNA genes, the simian virus 40 (SV40) enhancer, the histone H2B gene or the interferon-β gene to cite only a few [see detailed compilation in ref. 30]. In many cases the integrity of the octamer motif was shown to be essential for optimal expression of these genes. For example, an intact octamer sequence is required for the cell-cycle-modulated transcription of the histone H2B gene [31].

On the other hand, a number of experiments have shown that the octamer motif, although evidently not restricted to lymphoid-specific genes, is one of the most crucial elements for B-cell-specific immunoglobulin transcription. It is clearly the dominant element of the immunoglobulin promoter and its mutation or deletion is a strong down-mutation [28, 32–35]. The other conserved motifs of the immunoglobulin promoter like the heptamer motif (see fig. 2) appear to contribute only very modestly to activity [35–37]. The importance of the octamer site has also been extensively demonstrated in the context of the Igh enhancer [38–40].

Furthermore, insertion of an octamer site into a truncated β-globin [41, 42] or into a renin [43] promoter is sufficient to render the resulting promoters active preferentially in B cells, just as an immunoglobulin promoter is. Similarly, multimerized tandem copies of the octamer motif create an artificial enhancer which activates transcription of a β-globin test gene only in lymphoid cells, from a distance [44, 45] or even from a downstream position [39]. Hence, a variety of independent experiments provide evidence that the octamer motif can 'mediate lymphoid-specific transcription'. One thus faces the paradoxical situation that apparently the same DNA motif can be functionally active in all cells, or only in lymphoid (B) cells, depending on its context.

Octamer-Binding Factors. Part of this puzzle was solved when electrophoretic mobility shift assays revealed the presence of a number of nuclear proteins which specifically bind to the octamer sequence [39, 46–48]. Oct-1 (also called OTF-1, OBP 100, NF-A1 or NF-III [48–52]) has an apparent molecular weight of 95–100 kilodaltons and has been found in virtually all cells analyzed so far. It is therefore referred to as the ubiquitous octamer-binding factor. In vivo and in vitro experiments have shown that Oct-1 is responsible for the octamer-mediated activity associated with housekeeping genes [44, 49, 52, 53]. It was also shown that Oct-1 is necessary for efficient replication of adenovirus DNA, indicating that it may play a similar role in cellular DNA replication as well [51, 54]. Oct-2A (also called Oct-2, OTF-2 or NF-A2 [39, 47, 48, 55, 56]) has a lower apparent molecular weight (~60 kilodaltons) and is found in few cell types only, most notably in B cells and in some glial cells [see also ref. 57]. Oct-2B is another, less abundant lymphoid-specific factor of ~75 kilodaltons which is closely related to Oct-2A [58]. Collectively, Oct-2A/Oct-2B (referred to as Oct-2 for simplicity) comprise the lymphoid-specific octamer-binding factors. The finding that Oct-2 is already expressed in early pre-B cells [59, 60] (see fig. 1) is interesting and suggests that Oct-2 may be one of the factors which determine the B cell lineage [30].

Lymphoid-specific transcription mediated by the octamer site correlates tightly with the presence of Oct-2A/Oct-2B in B cells [39, 41, 61]. Furthermore, Oct-2 (OTF-2) purified from B cells stimulates specifically the transcription from an Igl promoter in a reconstituted in vitro transcription system [56].

Most conclusively, we have recently shown that B-cell-specific promoters containing an octamer site can be activated in non-B cells (HeLa) by cotransfection of an expression vector encoding Oct-2A [42], thus demonstrating that Oct-2A is indeed the factor involved in immunoglobulin promoter transcription. However, in the same assay, Oct-2A was not able to activate an enhancer consisting of multiple octamer sites [62], whereas such an enhancer was highly active in B cells, as mentioned above [39, 42, 44, 45]. To explain that result, we postulate that another related factor may be involved in enhancer activation (perhaps Oct-2B [58]), or that Oct-2A needs to act in concert with another yet unidentified B-cell-specific protein [62].

At this point the reader might wonder why an immunoglobulin promoter is not also active in non-B cells? Could it be that each octamer factor 'knows' where to bind, Oct-1 binding preferentially to ubiquitous octamer sites and Oct-2 to the lymphoid-specific sites (e.g. the immunoglobulin promoter octamer)? The answer is that there does not seem to be any preference in binding. In vitro, either factor binds equally well to either site [48, 52]. One thus has to look for an alternative explanation.

A model which accounts for the genetic and biochemical data obtained so far is presented in figure 3. An immunoglobulin promoter is proposed to be rather simple, with the octamer motif being, together with the TATA box, the most important element. The lymphoid-specific Oct-2A factor is a transcription activator which presumably can directly interact with the basic

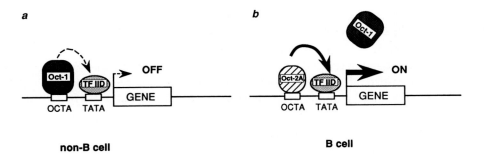

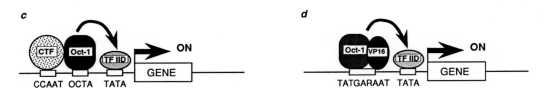

Fig. 3. Model for the differential regulation by the octamer motif [39, 41–44]. *a, b* An 'immunoglobulin-like' promoter which consists essentially of an octamer site and a TATA box is not activated in nonlymphoid cells since they only contain Oct-1 *(a)*. However, in B cells the same promoter combination is active as Oct-2A is present in addition to Oct-1 *(b)*. *c, d* The 'ubiquitous promoter' (e.g. the histone promoter) has a more complex structure and is readily active in non-B cells through cooperation between Oct-1 and other ancillary factors like the CAAT transcription factor, CTF [71] *(c)*. Oct-1 can also form a complex with the viral transactivator VP16 (and some cellular factors not indicated [65, 66, 69] *(d)*. This multiprotein complex activates transcription by binding an extended octamer-related site, such as -ATGCTAATGATATTC- (the so-called TATGARAAT motif), which is primarily found in promoters of herpes virus. The curved solid arrows indicate productive interaction between the upstream factors and the basic transcription machinery. This results in transcription, represented by the thick solid arrows. For simplicity RNA polymerase II is not indicated and TFIID symbolizes the TATA-box-binding factor as well as the other components of the basic transcription machinery.

transcription machinery and the TATA box factor (TFIID), and activate transcription. Transactivation experiments of minimal promoters with a cloned Oct-2A cDNA are compatible with that proposal [42].

On the other hand, the ubiquitous promoters which contain an octamer motif are usually complex and also contain other binding sites nearby (e.g. GC box, CAAT box, or 'proximal element' in the case of snRNA genes

[44, 53]). Mutagenesis experiments have shown that these sites as well as the octamer are essential for activity [69, 70]. Thus, Oct-1 is an activator which is necessary but not sufficient for the activity of these promoters. To be active, Oct-1 probably needs to cooperate with the transcription factors that bind to these elements (fig. 3).

An example of protein-protein interactions important for Oct-1 function is provided by the herpes virus protein VP16 (also called Vmw65). VP16 is a very strong transcription activator which has no DNA-binding activity by itself. However, together with Oct-1 (and some cellular proteins) VP16 can form a complex which binds to and activates transcription from an octamer-related site, the 'TATGARAAT' motif [44, 65–70] which is found primarily in herpesvirus promoters. The interaction is specific to Oct-1, and with Oct-2A the complex with VP16 forms only very poorly [66, 69].

In addition, several other less well characterized octamer-binding proteins have been identified in various tissues or cell lines. NF-A3 is found in malignant melanoma [72], Oct-T is a sea urchin testis-specific factor [73] and Oct4-Oct10 are present in mouse embryonic stem cells, oocytes or certain other tissues [74, 75]. Another octamer-binding factor (unfortunately also called NF-A3 [76]) is found in undifferentiated embryo carcinoma cells; interestingly, this factor disappears when the cells are induced to differentiate in vitro and was suggested by one group to be a repressor of enhancer function [76]. Another work, however, indicated that this protein is indeed a positive activator [72]. The cloning of this factor, which is underway, will certainly clarify this issue. Finally, two high-mobility group proteins have been found which preferentially bind A/T-rich sequences as well as the octamer site [77].

Octamer Factors: Homeobox Proteins. Recently, cDNAs for Oct-1 [78] and Oct-2A [42, 55, 79] have been isolated. Northern blot analysis of B cell RNA with an Oct-2A probe has revealed a complex family of hybridizing RNA species [42, 59] and some evidence of alternative splicing has been obtained [55]. This and the relatedness between Oct-2A and Oct-2B raises the possibility that these two octamer-binding factors could both be made from the same gene by differential splicing [58].

The deduced amino acid sequence of Oct-2A [42, 55, 79] shows a region of high homology to Oct-1, not surprisingly, as well as to another mammalian factor, Pit-1 [80, 81], and to a *C. elegans* gene product, Unc-86 [82]. This homology region has been called the POU domain (*P*it, *O*ct, *U*nc) and contains two conserved subregions: a POU-specific box and a POU-homeo

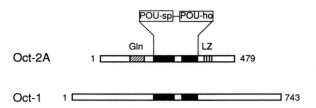

Fig. 4. Schematic structure of Oct-2A and Oct-1. The POU domain is represented by the solid boxes with the POU-specific (POU-sp) and the POU-homeo (POU-ho) subdomains indicated [83]. For Oct-2A, the leucine zipper is drawn (LZ), as well as a glutamine-rich region (Gln; hatched box) which is one of the transcription-activating regions of the protein [62, 88].

box which is about 30% homologous to homeoboxes found in many genes regulating development [83, 84] (fig. 4). Outside the POU domain these proteins are very different. Mutagenesis experiments suggest that the entire POU domain represents the DNA-binding domain of this class of proteins [55, 62, 85], although weak binding is still observed with the POU-homeo domain alone [86]. The POU-homeo domain of Oct-2A has been proposed to have a tri-α-helical structure similar to that of prokaryotic repressors and to be in direct contact with DNA [87]. Furthermore, it has been shown that Oct-2A contains several transcription-activating domains. One coincides with a glutamine-rich region and the other is localized in the C-terminal part of the protein [62, 88]. There may be another very weak activating region overlapping with the POU domain [62] (fig. 4).

The finding of a homeobox-related domain in Oct-2A provided the first demonstration of a specific function for a mammalian homeobox protein. For several of the previously identified *Drosophila* homeobox proteins it has been shown that the homeodomain is the DNA-binding domain of these proteins [84]. The target sequences to which these proteins bind are typically A/T-rich [89, 90] and contain some loose homologies to the octamer site [79, 91]. In agreement with that finding it was shown that Oct-2A can specifically bind to such target sequences in vitro [79, 92] and even activate transcription from there in vivo [91]. Conversely, two *Drosophila* homeobox proteins, Ubx and Abdr, can activate transcription in mammalian cells from the natural *Drosophila* site or from the octamer site [91]. Hence, the conservation of the homeodomain between distantly related proteins appears to determine a similar DNA-binding specificity, and selective interactions with cellular or viral factors are likely to modulate that specificity.

Furthermore, Oct-2A also contains a region with homology to the 'leucine zipper', a protein dimerization motif found in several DNA-binding proteins [93] (fig. 2). However, this region appears to be largely dispensable for binding to DNA [55, 62] or even transcription activation [62, 88].

In closing, a further demonstration of the role of the octamer motif and of Oct-2 for immunoglobulin expression comes from cell fusion experiments. When immunoglobulin-expressing B cells are fused with cells of a different kind (e.g. fibroblasts), immunoglobulin expression is rapidly suppressed at the transcriptional level [94], a phenomenon called extinction [94, 95]. Both the κ-promoter and the Igh enhancer are targets for this down-regulation in a hybrid environment and can confer it individually upon a heterologous test gene [94, 95]. We have recently demonstrated that the octamer site of the κ-promoter is essential for mediating extinction. Simple replacement of the κ-promoter octamer site with a binding site for a ubiquitous factor (e.g. Sp1) is sufficient to bypass extinction. In these hybrids between fibroblasts and B cells, extinction correlates with suppression of *oct-2* gene expression, as shown by the lack of Oct-2 protein and of the corresponding RNAs [96]. Moreover, in such hybrids, transcription form the κ-promoter can be at least partly restored by introduction of an expression vector encoding Oct-2A [96]. Thus, in that system extinction of immunoglobulin expression is reversible and correlates with the lack of a necessary cell-specific transcription factor, Oct-2A.

Ephrussi Boxes

The E boxes (consensus: -CAGGTGGC-) have been originally identified in the Igh enhancer on the basis of in vivo footprints which were detected specifically in B cells [97, 98]. Subsequent in vitro studies with nuclear extracts showed that individual protein bind to three E sites in the Igh enhancer and to two related sites in the κ-enhancer [40, 99–101]. E-box-related motifs have also been described in the insulin [102] and in the muscle creatine kinase [103] enhancers.

Paradoxically, although the in vivo footprints on the Igh enhancer E sites were B-cell-specific [97, 98], the proteins binding these sites in vitro were also found in nuclear extracts from nonlymphoid cells and are probably ubiquitous [38, 100, 101, 104]. Mutational analysis of the Igh and κ-enhancer E sites demonstrated that they are indeed important for immunoglobulin gene expression, particularly when several sites are mutated in combination [38, 40, 102]. However, the effects of the mutations were not restricted to B cells [38, 40], suggesting that the factors binding to the E boxes can probably contribute to transcription activation in many different cell types. This argument is further supported by the finding that mutation of E sites of either the insulin or the muscle creatine kinase enhancer has a detrimental effect in pancreatic or muscle cells [102, 103].

In a similar experimental approach to what had been done with the octamer motif [41–43], we tested the activity of the Igh enhancer E boxes by inserting them individually next to a β-globin TATA box. We found that in this assay system a single copy of an E3 box can activate transcription efficiently both in B and non-B cells [105]. Thus, in the natural situation, differential accessibility of the sites (e.g. chromatin structure) or modification of the factors (see below) may account for the B-cell-specific binding to these sites.

The recent cloning of a cDNA encoding a κE2-site-binding protein (called E47) revealed that this gene is expressed in various cell types, as expected from the ubiquitous distribution of the protein [106]. So far there is no direct evidence that E47 is indeed a transcription activator. However, E47 exemplifies a new class of DNA-binding proteins which show some very interesting biochemical properties. The predicted amino acid sequence of E47 [106] showed a region of homology to a group of proteins whose relationship to one another had been noted previously [107]. Among these proteins are: several *Drosophila* gene products, like daughterless [108], achaete-scute [109] and twist [110]; MyoD, a factor controlling myogenic differentiation [107], and myc proteins, like L-myc, c-myc and N-myc [111]. Also some relationship to lamins can be evidenced [106].

The homology region can be predicted to contain two amphipathic α-helices separated by a loop, giving a structure resembling a Greek Ω [106, 112]. This domain has been called the helix-loop-helix (HLH) motif. A region rich in basic amino acid residues is found N-terminal of the HLH domain in E47 [106].

Experiments with in vitro translated proteins showed that E47 binds to the κE2 site as a dimer, and that the HLH domain is essential for binding and dimerization [106]. Most interestingly, in vitro, E47 can readily form heterodimers with other proteins of the HLH class, such as achaete-scute and MyoD, and bind with high affinity and specificity to the κE2 site [113]. Although it is not known at present whether such heterodimers are also found in vivo, the hypothesis of combinatorial heterodimerization is very attractive to explain how cell-specific responses could be generated from ubiquitous components. For example, one could imagine that certain heterodimers will have an altered (e.g. increased) DNA-binding activity or transcription activation potential. In the simplest model the relative amounts of the different partners would determine which heterodimers are present in a particular cell type.

Finally, the gene encoding E47 appears also to be involved in certain pathological conditions. In about 30% of pediatric pre-B cell acute lymphocytic leukemias a chromosome translocation can be evidenced which splits the E47 gene [114]. The translocation leads to the creation of a hybrid gene between the E47 gene 5'-end and a gene on chromosome 1 [115]. Cloning and characterization of the hybrid cDNA derived from a cell line having the chromosomal translocation showed that it encodes a protein lacking the DNA-binding and dimerization domains of E47, and having instead a homeobox-related sequence [115]. A speculative model trying to tie these findings together suggests that the homeodomain appended to E47 provides a new DNA-binding specificity which leads to inappropriate activation (or repression) of cellular target genes and oncogenesis [115].

κB Site

The κB site (-G/AGGGACTTTCC-) has first been identified as a motif exactly conserved between the κ- and SV40 enhancers, and bound in vitro by a nuclear factor which has been dubbed NF-κB [99]. In vivo experiments showed that the κB site is crucial for κ-enhancer activity [40, 116].

At first it appeared that NF-κB was restricted to B cells, because it was found constitutively only in B cells of the appropriate stage for Igl expression [117] (fig. 1). However, it was soon realized that the same nuclear-binding activity could be induced in pre-B cells by treatment with the mitogen bacterial lipopolysaccharide (LPS; fig. 1) or even in non-B cells by treatment with phorbol esters such as 12-O-tetradecanoylphorbol-13-acetate (TPA) [118]. Furthermore, other inducers of NF-κB have also been identified such as virus infection or treatment of cells with double-stranded RNA [119, 120], and it now seems that NF-κB is a virtually ubiquitous protein whose DNA-binding activity can be induced by a variety of different signals [117]. In accordance with these findings, in transient transfection assays, the κB site behaves as an inducible enhancer under conditions which activate NF-κB binding [60, 121]. The mechanism of NF-κB induction has been analyzed in great detail. Induction of NF-κB can be observed in the absence of protein synthesis, implying that it involves modification of a preexising precursor [118].

By treating extracts with dissociating agents such as formamide or deoxycholate, the presence of NF-κB binding activity could be unmasked in

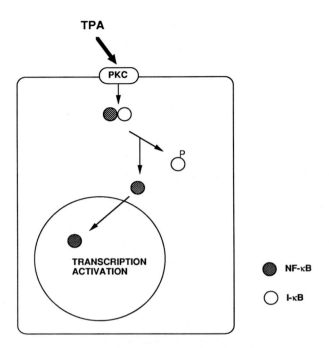

Fig. 5. A model for signal transduction by NF-κB [116, 120, 121]. As an example of an inducer, TPA activates protein kinase C (PKC) which, in turn, phosphorylates I-κB. This leads to dissociation of the NF-κB I-κB complex and subsequent translocation of NF-κB into the nucleus where it can bind DNA and activate transcription. Activation of NF-κB by other inducers may use different routes.

the cytoplasm from non-activated cells [122]. Moreover, a protein of about 68 kilodaltons was identified, which is able to specifically inhibit the DNA-binding activity of NF-κB [123]. The protein has been called inhibitor-κB (I-κB) and can only be evidenced by dissociating treatment of cytoplasmic extracts, indicating that there is no excess of free I-κB in the ground state [123]. It seems likely that activation involves modification of I-κB and not of NF-κB, because I-κB is able to inhibit NF-κB irrespective of its source [123]. Since NF-κB can be activated in vivo by treatment with TPA, a known activator of protein kinase C, it has been proposed that phosphorylation of I-κB is the critical step controlling dissociation of the complex and subsequent nuclear translocation of NF-κB [117, 122, 123] (fig. 5). Very recent experiments support that hypothesis and show that I-κB is indeed a target for phosphorylation [124]. Within the B cell lineage, the NF-κB system can be viewed as a developmental molecular switch. In pre-B cells, NF-κB is in a dormant, inducible form. Around the pre-B-B cell transition, a molecular event takes place which renders NF-κB constitutively nuclear (fig. 1). Trying to understand what is the basis for that apparently permanent change is a fascinating issue.

κB sites are not restricted to the κ- and SV40 enhancers, but are also found in the control regions of several cellular genes (e.g. interleukin-2, interleukin-2 receptor α-chain, interferon-β) and viruses, most notably in the human cytomegalovirus and immunodeficiency virus (HIV) [reviewed in ref. 117]. The finding of κB sites in the HIV long terminal repeat (LTR) has drawn particular attention, since agents which lead to T cell activation also result in activation of NF-κB [99, 125]. In agreement with this it was shown that the κB sites are essential for transcription from the HIV LTR in activated T cells [125]. Thus, the activation of NF-κB concomitant to T cell activation could be directly responsible for increased transcription of the HIV provirus and result in elevated virus production [125].

Other, unrelated transcription factors also bind to the κB site, thus complicating the interpretation of some results. For example, H2TF1 is a protein which recognizes a κB-like site in a mouse class I histocompatibility gene [126]. This factor, which is cloned [127], is a constitutively active nuclear protein found in many cell types and is clearly distinct from NF-κB.

Unfortunately, the very low abundance of NF-κB has made its biochemical purification very difficult and the cloning of its gene has not been achieved yet.

Conclusions and Perspectives

Our understanding of the regulation of immunoglobulin expression (and of gene expression in general) has progressed enormously with the identification and in some cases cloning of the factors involved. However, this knowledge relies mostly on the study of individual model systems (e.g. octamer, E boxes). We do not understand yet how the various elements 'fit together'. For example, studies on the SV40 enhancer have shown that such a structure has a very high degree of hierarchical organization which may be overlooked by studying individual elements [41]. In practice, this means that certain elements will reveal their 'true role' (e.g. cooperativity between two factors binding to adjacent motifs) only when in the proper context.

Also, it will be interesting to see whether transcription regulation is perhaps, under some circumstances, coordinated with other levels of regulation like, e.g. RNA processing or DNA rearrangement. For instance, there is ample evidence that DNA rearrangement and 'accessibility' of the immunoglobulin locus correlate [2, 4]. Is this merely guilt by association, or is there a causality relationship? Is one phenomenon inducing the other?

How does the immune system of a mouse lacking Oct-2 or E47 develop? Is, as expected, the transcription of immunoglobulin genes impaired? These and many other questions can now be addressed, as the techniques for homologous recombination in embryonic stem cells are coming of age and should allow the replacement of normal cellular DNA sequences in the mouse germ line by mutant alleles generated in vitro [128]. With no doubt this will be a very powerful tool to study all aspects of gene regulation, and such experiments have been initiated in several laboratories. Having transcription factors purified in large amounts (from natural sources or from molecular clones) now allows biochemical studies on the mechanism of transcription regulation. Which factors are directly interacting e.g. with one another (see fig. 3), with RNA polymerase II?

Also, and perhaps more importantly for the biologists, it is now possible to study the regulation of the genes encoding these factors. For example, what is the primary level of control of the *oct-2* gene; is it transcriptional or perhaps posttranscriptional control? What physiological stimuli affect the expression of such a gene? Is this gene autoregulated, like many other homeobox genes [84]? I have previously mentioned that LPS induces accumulation of *oct-2* RNA and protein [59, 60] and that fusion of B cells with fibroblasts results in a switch off of the *oct-2* gene. In both cases, which is the mechanism involved?

Last, but not least, the identification of additional regulatory elements in the immunoglobulin loci is an important task. On the basis of cell lines which had deleted the intron Igh enhancer and yet retained high levels of immunoglobulin expression, it had been argued that the Igh enhancer is only required for establishment, but not for maintenance of immunoglobulin transcription [129, 130]. Now, however, there is increasing evidence that this is probably not the correct explanation [131, 132], and that, rather, additional regulatory elements have to be identified. Are there additional enhancers in the Igh locus, spread over the C regions (see fig. 2)? If there are, could these elements be involved in regulating class switch recombination (which also correlates with accessibility of the locus [133])? The recent finding of an enhancer 3' of the Cκ region [17] and also 25 kb downstream of the Cα gene [18] gives further support to that hypothesis. Also, what element (and factor) is controlling the transient transcription of the V region promoters before they are rearranged [13]?

Surely the study of the regulation of immunoglobulin expression will continue to provide a fascinating field of study.

References

1 Tonegawa S: Somatic generation of antibody diversity. Nature 1983;302:575–581.
2 Alt FW, Blackwell TK, Yancopoulos GD: Development of the primary antibody repertoire. Science 1987;238:1079–1087.
3 Sakano H, Maki R, Kurosawa Y, et al: Two types of somatic recombination sites are necessary for the generation of a complete immunoglobulin heavy chain gene. Nature 1980;286:676–683.
4 Schlissel MS, Baltimore D: Activation of immunoglobulin gene rearrangement correlates with induction of germline kappa gene transcription. Cell 1986;58:1001–1007.
5 Yancopoulos G, Blackwell TK, Suh H, et al: Introduced T cell receptor variable region gene segments recombine in pre-B cells: Evidence that B and T cells use a common recombinase. Cell 1986;44:251–259.
6 Schatz DG, Oettinger MA, Baltimore D: The V(D)J recombination activating gene, RAG-1. Cell 1989;59:1035–1048.
7 Galli G, Guise JW, Tucker PW: Poly(A) site choice rather than splice site choice governs the regulated production of IgM heavy chain RNAs. Proc Natl Acad Sci USA 1988;85:2439–2443.
8 Galli G, Guise JW, McDevitt MA, et al: Relative position and strengths of poly(A) sites as well as transcription termination are critical to membrane versus secreted μ-chain expression during B cell development. Genes Dev 1987;1:471–481.
9 Peterson ML, Perry RP: Regulated production of μm and μs mRNA requires linkage of the poly(A) addition sites and is dependent on the length of the μm-μs intron. Proc Natl Acad Sci USA 1986;83:8883–8887.
10 Mather EL, Nelson KJ, Haimovitch J, Perry R: Mode of regulation of immunoglobulin μ- and δ-chain expression varies during B-lymphocyte maturation. Cell 1979;36:329–338.
11 Gerster T, Picard D, Schaffner W: During B cell differentiation enhancer activity and transcription rate of immunoglobulin heavy chain genes are high before mRNA accumulation. Cell 1986;45:45–52.
12 Rajewsky K, Förster I, Cumano A: Evolutionary and somatic selection of the antibody repertoire in the mouse. Science 1987;238:1088–1094.
13 Yancopoulos GD, Alt FW: Developmentally controlled and tissue-specific expression of unrearranged V_H gene segments. Cell 1985;40:271–281.
14 Banerji J, Olson L, Schaffner W: A lymphocyte-specific cellular enhancer is located downstream of the joining region in immunoglobulin heavy chain genes. Cell 1983;33:729–740.

15 Gillies SD, Morrison SL, Oi VT, Tonegawa S: A tissue-specific transcription enhancer element is located in the major intron of the rearranged immunoglobulin heavy chain gene. Cell 1983;33:717–728.
16 Neuberger MS: Expression and regulation of immunoglobulin heavy chain gene transfected into lymphoid cells. EMBO J 1983;2:597–600.
17 Meyer KB, Neuberger MS: The immunoglobulin κ locus contains a second, stronger B cell specific enhancer which is located downstream of the constant region. EMBO J 1989;8:1959–1964.
18 Petterssen S, Cook GP, Brüggemann M, Williams GT, Neuberger MS: A second B cell specific enhancer 3′ of the immunoglobulin heavy chain locus. Nature 1990;344:165–168.
19 Picard D, Schaffner W: A lymphocyte-specific enhancer in the mouse immunoglobulin κ gene. Nature 1984;322:656–659.
20 Picard D, Schaffner W: Cell-type preference of immunoglobulin kappa and lambda gene promoters. EMBO J 1985;4:2831–2838.
21 Grosschedl R, Baltimore D: Cell type specificity of immunoglobulin gene expression is regulated by at least three DNA sequence elements. Cell 1985;41:885–897.
22 Hatzopoulos AK, Schlokat U, Gruss P: Enhancers and other cis-regulatory sequences; in Hames BD, Glover DM (eds): Eucaryotic RNA Synthesis and Processing. Horizons in Molecular Biology. Oxford, IRL Press, 1988, pp 43–96.
23 Matthias P, Gerster T, Bohmann D, Schaffner W: Cell-type specificity of transcription: The immunoglobulin heavy chain enhancer as a model system; in Eckstein F, Lilley DMJ (eds): Nucleic Acids and Molecular Biology. Berlin, Springer, 1987, vol 1, pp 221–240.
24 Wasylyk B, Wasylyk C: The immunoglobulin heavy chain B lymphocyte enhancer efficiently stimulates transcription in nonlymphoid cells. EMBO J 1986;5:553–560.
25 Imler JC, Lemaire C, Wasylyk C, Wasylyk B: Negative regulation contributes to tissue specificity of the immunoglobulin heavy chain enhancer. Mol Cell Biol 1987;7:2558–2567.
26 Kadesch T, Zervos P, Ruezinsky D: Functional analysis of the murine IgH enhancer: Evidence for negative control of cell-type specificity. Nucleic Acids Res 1986;14:8209–8221.
27 Scheuerman RH, Chen U: A developmental-specific factor binds to suppressor sites flanking the immunoglobulin heavy chain enhancer. Genes Dev 1989;3:1255–1266.
28 Falkner FG, Zachau HG: Correct transcription of an immunoglobulin κ gene requires an upstream fragment containing conserved sequence elements. Nature 1984;310:71–74.
29 Parslow TG, Blair DL, Murphy WJ, Granner DK: Structure of the 5′ ends of immunoglobulin genes: A novel conserved sequence. Proc Natl Acad Sci USA 1984;81:2650–2654.
30 Schreiber E, Müller MM, Schaffner W, Matthias P: Octamer transcription factors mediate B-cell specific expression of immunoglobulin heavy chain genes; in Renkawitz R (ed): Tissue Specific Gene Expression. Weinheim, VCH Publishers, 1989, pp 33–54.
31 LaBella F, Sive HL, Roeder RG, Heintz N: Cell-cycle regulation of a human histone H2B gene is mediated by the H2B subtype-specific consensus element. Genes Dev 1988;2:32–39.
32 Mason JO, Williams GT, Neuberger MS: Transcription cell-type specificity is controlled by an immunoglobulin V_H promoter that includes a functional consensus sequence. Cell 1985;41:479–487.

33 Bergeman Y, Rice D, Grosschedl R, Baltimore D: Two regulatory elements for immunoglobulin κ light chain expression. Proc Natl Acad Sci USA 1984;81:7041–7045.

34 Ballard DW, Bothwell A: Mutational analysis of the immunoglobulin heavy chain promoter region. Proc Natl Acad Sci USA 1986;83:9626–9630.

35 Eaton S, Calame K: Multiple DNA sequence elements are necessary for the function of an immunoglobulin heavy chain promoter. Proc Natl Acad Sci USA 1987;84:7634–7638.

36 Pollinger L, Yoza BK, Roeder RG: Functional cooperativity between protein molecules bound at two distinct sequence elements of the immunoglobulin heavy chain promoter. Nature 1989;337:573–576.

37 Kemler I, Schreiber E, Müller MM, Matthias P, Schaffner W: Octamer transcription factors bind to two different sequence motifs of the immunoglobulin heavy chain promoter. EMBO J 1989;8:2001–2008.

38 Perez-Mutul J, Macchi M, Wasylyk B: Mutational analysis of the contribution of sequence motifs within the IgH enhancer to tissue specific transcriptional activation. Nucleic Acids Res 1988;16:6085–6096.

39 Gerster T, Matthias P, Thali M, Jiricny J, Schaffner W: Cell type-specific elements of the immunoglobulin heavy chain gene enhancer. EMBO J 1987;6:1323–1330.

40 Lenardo MJ, Pierce J, Baltimore D: Protein binding sites in Ig enhancers determine transcriptional activity and inducibility. Science 1987;236:1573–1577.

41 Wirth T, Staudt, L, Baltimore D: An octamer oligonucleotide upstream of a TATA motif is sufficient for lymphoid-specific promoter activity. Nature 1987;329:174–177.

42 Müller MM, Ruppert S, Schaffner W, Matthias P: A cloned octamer transcription factor stimulates transcription from lymphoid-specific promoters in non-B cells. Nature 1988;336:544–551.

43 Dreyfus M, Doyen N, Rougeon F: The conserved decanucleotide from the immunoglobulin heavy chain promoter induces a very high transcriptional activity in B-cells when introduced into a heterologous promoter. EMBO J 1987;6:1685–1690.

44 Tanaka M, Grossniklaus U, Herr W, Hernandez N: Activation of the U2 snRNA promoter by the octamer motif defines a new class of RNA polymerase II enhancer elements. Genes Dev 1988;2:1764–1778.

45 Fromenthal C, Kanno M, Nomiyama H, Chambon P: Cooperativity and hierarchical levels of functional organization in the SV40 enhancer. Cell 1988;54:943–953.

46 Singh H, Sen R, Baltimore D, Sharp PA: A nuclear factor that binds to conserved sequence motif in transcription control elements of immunoglobulin genes. Nature 1986;319:154–158.

47 Landolfi NF, Capra DJ, Tucker PW: Interaction of cell-specific nuclear proteins with immunoglobulin V_H promoter region sequences. Nature 1986;323:548–551.

48 Staudt LM, Singh H, Sen R, Sharp PA, Baltimore D: A lymphoid-specific protein binding to the octamer motif of immunoglobulin genes. Nature 1986;323:640–643.

49 Fletcher C, Heintz N, Roeder RG: Purification and characterization of OTF-1, a transcription factor regulating cell cycle expression of a human histone H2B gene. Cell 1987;51:773–781.

50 Sturm R, Baumruker T, Franza BR, Herr W: A 100 KDa HeLa cell octamer binding protein (OBP100) interacts differently with two separate octamer-related sequences within the SV40 enhancer. Genes Dev 1987;1:1147–1160.

51 Pruijn GM, van Driel W, van der Vliet PC: Nuclear factor III: A novel sequence-specific DNA binding protein from HeLa cells stimulating adenovirus replication. Nature 1986;322:656–659.

52 Sive HL, Roeder RG: The interaction of a common factor with conserved promoter and enhancer sequences in histone H2B, immunoglobulin and U2 small nuclear RNA genes. Proc Natl Acad Sci USA 1986;83:6382–6386.

53 Schaffner W: How do different transcription factors binding the same DNA sequence sort out their jobs? Trends Genet 1989;5:37–39.

54 O'Neill EA, Fletcher C, Burrow CR, Heintz N, Roeder RG, Kelly TJ: Transcription factor OTF-1 is functionally identical to the DNA replication factor NFIII. Science 1988;241:1210–1213.

55 Clerc RG, Corcoran LM, LeBowitz JH, Baltimore D, Sharp PA: The B cell specific Oct-2 protein contains POU-box and homeo-box type domains. Genes Dev 1988;2:1570–1581.

56 Scheidereit C, Heguy A, Roeder RG: Identification and purification of a human lymphoid-specific octamer-binding protein (OTF-2) that activates transcription of an immunoglobulin promoter in vitro. Cell 1987;51:783–793.

57 He X, Treacy MN, Simmons DM, Ingraham HA, Swanson L, Rosenfeld MG: Expression of a large family of POU-domain regulatory genes in mammalian brain development. Nature 1989;340:35–42.

58 Schreiber E, Matthias P, Müller MM, Schaffner W: Identification of a novel lymphoid specific octamer binding protein (OTF-2B) by proteolytic clipping band-shift assay (PCBA). EMBO J 1988;7:4421–4229.

59 Staudt LM, Clerc RG, Singh H, Sharp PA, Baltimore D: Cloning of a cDNA encoding a B-cell restricted octamer binding factor. Science 1988;241:577–580.

60 Wirth T, Baltimore D: Nuclear factor NF-κB can interact fuctionally with its cognate site to provide lymphoid-specific promoter function EMBO J 1988;7:3109–3113.

61 Mishizuma-Sugano J, Roeder RG: Cell type specific transcription of an immuno-globulin κ light chain in vitro. Proc Natl Acad Sci USA 1986;83:8511–8515.

62 Müller MM, Schaffner W, Matthias P: Transcription factor Oct-2A contains func-tionally redundant activating domains and works selectively from a promoter but not from a remote enhancer position in non-lymphoid cells. EMBO J, in press.

63 Cox PM, Temperley SM, Kumar H, Goding CR: A distinct octamer-binding pro-tein present in malignant melanoma cells. Nucleic Acids Res 1988;16:11047–11055.

64 Barberis A, Superti-Furga G, Busslinger M: A CAAT displacement factor binds in the promoter of a sea urchin histone gene. Cell 1987;50:347–359.

65 Schöler HR, Balling R, Hatzopoulos HA, Suzuki N, Gruss P: Octamer binding proteins confer transcriptional activity in early mouse embryogenesis. EMBO J 1989;9:2551–2557.

66 Schöler HR, Hatzopoulos AK, Balling R, Gruss P: A family of octamer-specific proteins present during mouse embryogenesis: Evidence for germline-specific ex-pression of an Oct factor. EMBO J 1989;9:2543–2550.

67 Lenardo MJ, Staudt L, Robbins P, Kuang A, Mulligan R, Baltimore D: Repression of the IgH enhancer in teratocarcinoma cells associated with a novel octamer factor. Science 1989;243:544–546.

68 Eckner R, Birnstiel ML: Cloning of cDNAs coding for HMGI and HMGY proteins: Both are capable of binding the octamer sequence motif. Nucleic Acids Res 1989;17:5947–5959.

69 Sive HL, Heintz N, Roeder RG: Multiple sequence elements required for maximal in vitro transcription of a human histone H2B gene. Mol Cell Biol 1986;6:3329–3340.

70 Ares M, Chung JS, Giglio L, Weiner AM: Distinct factors with Sp1 and NF-A specificities bind to adjacent functional elements of the human U2 snRNA gene enhancer. Genes Dev 1987;1:808–817.

71 Santoro C, Mermod N, Andrews PC, Tjian R: A family of human CCAAT-box-binding proteins active in transcription and DNA replication: cloning and expression of multiple cDNAs. Nature 1988;334:218–224.

72 Gerster T, Roeder RG: A herpesvirus trans-activating protein interacts with transcription factor OTF-1 and other cellular proteins. Proc Natl Acad Sci USA 1988;85:6347–6351.

73 Stern S, Tanaka M, Herr W: The Oct-1 homeodomain directs formation of a multiprotein-DNA complex with the HSV transactivator VP16. Nature 1989;341:624–630.

74 O'Hare P, Goding CR, Haigh A: Direct combinatorial interaction between a herpes simplex virus regulatory protein and a cellular octamer-binding factor mediates specific induction of virus immediate-early gene expression. EMBO J 1988;7:4231–4238.

75 O'Hare P, Goding CR: Herpes virus regulatory elements and the immunoglobulin octamer bind a common factor and are both targets for virion transactivation. Cell 1988;52:435–445.

76 Kristie TM, LeBowitz JH, Sharp PA: The octamer-binding proteins form multiprotein-DNA complexes with the HSV α-TIF regulatory protein. EMBO J 1989;8:4229–4238.

77 Robertson M: Homeo boxes, POU proteins and the limits to promiscuity. Nature 1988;336:522–524.

78 Sturm RA, Das G, Herr W: The ubiquitous octamer binding protein Oct-1 contains a POU domain with a homeo box subdomain. Genes Dev 1988;2:1582–1599.

79 Scheidereit C, Cromlish JA, Kawakami K, Gerster T, Roeder RG: A human lymphoid-specific transcription that activates immunoglobulin genes is a homeobox protein. Nature 1988;336:551–557.

80 Ingraham HA, Chen R, Mangalam HP, Elsholtz HP, Flynn SE, Lin CR, Simmons DM, Swanson LW, Rosenfeld MG: A tissue specific transcription factor containing a homeo domain specifies a pituitary phenotype. Cell 1988;55:519–529.

81 Bodner M, Castrillo JL, Theil LE, Karin M: The pituitary-specific transcription factor GHF-1 is a homeobox protein. Cell 1988;55:505–518.

82 Finney M, Ruvkun G, Horvitz HR: The C. elegans cell lineage and differentiation gene unc-86 encodes a protein containing a homeo domain and extended sequence similarity to mammalian transcription factors. Cell 1988;55:757–769.

83 Herr W, Sturm R, Clerc RG, Baltimore D, Sharp PA, Ingraham HA, Rosenfeld MG, Finney M, Ruvkun G, Horvitz HR: The POU domain: A large conserved region in the mammalian pit-1, oct-1, oct-2, and C. elegans unc-86 gene products. Genes Dev 1988;2:1513–1516.

84 Scott MP, Tamkun JW, Hartzell GW: The struture and function of the homeodomain. BBA Rev Cancer 1989;989:25–48.

85 Sturm RA, Herr W: The POU domain is a bipartite DNA-binding structure. Nature 1988;336:601–604.

86 Theils LE, Castrillo JL, Wu D, Karin M: Dissection of functional domains of the pituitary-specific transcription factor GHF-1. Nature 1989;342:945–948.

87 Garcia-Blanco MA, Clerc RG, Sharp PA: The DNA binding homeo domain of the Oct-2 protein. Genes Dev 1989;3:739–745.
88 Desplan C, Theis, J O'Farrel P: The sequence specificity of homeodomain DNA interaction. Cell 1988;54:1081–1090.
89 Hoey T, Levine M: Divergent homeobox proteins recognize similar DNA sequences in Drosophila. Nature 1988;332:858–861.
90 Thali M, Müller MM, DeLorenzi M, Matthias P, Bienz M: Drosophila homeotic genes encode transcription activators similar to mammalian OTF-2. Nature 1988; 336:598–601.
91 Ko HS, Fast P, McBride W, Staudt LM: A human protein specific for immunoglobulin octamer DNA motif contains a functional homeobox domain. Cell 1988;55:135–144.
92 Landschulz WH, Johnson PF, McKnight SL: The leucine zipper: A hypothetical structure common to a new class of DNA binding proteins. Science 1988;240:1759–1764.
93 Tanaka M, Herr W: Differential transcriptional activation by Oct-1 and Oct-2: Interdependent activation domains induce Oct-2 phosphorylation. Cell, in press.
94 Junker S, Nielsen V, Matthias P, Picard D: Both immunoglobulin promoter and enhancer sequences are targets for suppression in myeloma-fibroblast hybrid cells. EMBO J 1988;7:3093–3098.
95 Zaller D, Yu H, Eckhardt L: Genes activated in the presence of an immunoglobulin enhancer or promoter are negatively regulated by a T lymphoma cell line. Mol Cell Biol 1988;8:1932–1939.
96 Junker S, Pedersen S, Schreiber E, Matthias P: Extinction of immunoglobulin expression in somatic cell hybrids: Involvement of the octamer factor. Cell, in press.
97 Ephrussi A, Church GM, Tonegawa S, Gilbert W: B-lineage-specific interactions of an immunoglobulin enhancer with cellular factors in vivo. Science 1985;227:134–140.
98 Church GM, Ephrussi A, Gilbert W, Tonegawa S: Cell type specific contacts to immunoglobulin enhancers in nuclei. Nature 1985;313:798–801.
99 Sen R, Baltimore D: Multiple nuclear factors interact with the immunoglobulin enhancer sequence. Cell 1986;46:705–716.
100 Augereau P, Chambon P: The mouse immunoglobulin heavy chain enhancer: Effect on transcription in vitro and binding of proteins present in HeLa and lymphoid B cell extracts. EMBO J 1986;5:1791–1797.
101 Schlokat U, Bohmann D, Schöler H, Gruss P: Nuclear factors binding specific sequence within the immunoglobulin enhancer interact differentially with other enhancer elements. EMBO J 1986;5:3251–3258.
102 Moss LG, Moss JB, Rutter WJ: Systematic binding analysis of the insulin gene transcription control region: Insulin and immunoglobulin enhancers utilize similar transactivators. Mol Cell Biol 1988;8:2620–2627.
103 Buskin JN, Hauschka SD: Identification of a myocyte nuclear factor which binds to the muscle-specific enhancer of the mouse muscle creatine kinase gene. Mol Cell Biol 1989;9:2627–2640.
104 Peterson CL, Calame K: Complex protein binding within the mouse immunoglobulin heavy chain enhancer. Mol Cell Biol 1987;7:4194–4200.
105 Elmaleh N, Matthias P, Schaffner W: A factor known to bind to the endogenous IgH enhancer only in lymphocytes is a ubiquitously active transcription factor. Eur J Biochem, in press.

106 Murre C, Schonleber-McCaw P, Baltimore D: A new DNA binding and dimerization motif in immunoglobulin enhancer binding, daughterless, MyoD, and myc proteins. Cell 1989;56:777–783.

107 Davis RL, Weintraub H, Lassar AB: Expression of a single transfected cDNA converts fibroblasts to myoblasts. Cell 1987;51:987–1000.

108 Caudy M, Vassin H, Brand M, Tuma R, Jan LY, Jan YN: Daughterless, a Drosophila gene essential for both neurogenesis and sex determination has sequence similarities to myc and the achaete-scute complex. Cell 1988;55:1061–1067.

109 Villares R, Cabrera CV: The achaete-scute gene complex of *D. melanogaster:* Conserved domains in a subset of genes required for neurogenesis and their homology to myc. Cell 1987;50:415–424.

110 Thisse B, Stoetzel C, Gorostiza-Thisse C, Perrin-Schmitt F: Sequence of the twist gene and nuclear localization of its protein in endomesodermal cells of early Drosophila embryos. EMBO J 1988;7:2175–2183.

111 DePinho RA, Hatton KS, Tesfaye A, Yancopoulos GD, Alt FW: The human myc gene family: Structure and activity of L-myc and an L-myc pseudogen. Genes Dev 1987;1:1311–1326.

112 Lesczinsky JF, Rose GD: Loops in globular proteins: A novel category of secondary structure. Science 1986;234:849–855.

113 Murre C, Schonleber-McCaw P, Vassin H, Caudy M, Jan LY, Jan YN, Cabrera CV, Buskin JN, Hauschka SD, Lassar AB, Weintraub H, Baltimore D: Interactions between heterologous helix-loop-helix proteins generate complexes that bind specifically to common DNA sequences. Cell 1989;58:537–544.

114 Mellentin JD, Murre C, Donlon TA, McCaw PS, Smith SD, Carroll AJ, McDonald ME, Baltimore D, Cleary ML: The gene for enhancer binding proteins E12/E47 lies at the t(1;19) breakpoint in acute leukemias. Science 1989;246:379–382.

115 Kamps MP, Murre C, Sun X, Baltimore D: A new homeobox gene contributes the DNA-binding domain of the t(1;19) translocation protein in pre-B ALL. Cell 1990; 60:547–555.

116 Atchinson ML, Perry RP: The role of the κ enhancer and its binding factor NF-κB in the developmental regulation of κ gene transcription. Cell 1987;48:121–128.

117 Lenardo MJ, Baltimore D: Nf-κB A pleiotropic mediator of inducible and tissue-specific gene control. Cell 1989;58:227–229.

118 Sen R, Baltimore D: Inducibility of κ immunoglobulin enhancer binding protein NF-κB by a posttranslational mechanism. Cell 1986;47:921–928.

119 Visvanathan KV, Goodbourne S: Double-stranded RNA activates binding of NF-κB to an inducible element of the human β-interferon promoter. EMBO J 1989;8:1129–1138.

120 Lenardo MJ, Fan CM, Maniatis T, Baltimore D: The involvement of NF-κB in β-interferon gene regulation reveals its role as a widely inducible second messenger. Cell 1989;57:287–294.

121 Pierce JW, Lenardo MJ, Baltimore D: An oligonucleotide that binds nuclear factor NF-κB acts as a lymphoid-specific and inducible enhancer element. Proc Natl Acad Sci USA 1988;85:1482–1486.

122 Baeuerle PA, Baltimore D: Activation of DNA-binding activity in an apparent cytoplasmic precursor of the NF-κB transcription factor. Cell 1988;53:211–217.

123 Baeuerle PA, Baltimore D: IκB: A specific inhibitor of the NF-κB transcription factor. Science 1988;242:540–546.

124 Gosh S, Baltimore D: Activation in vitro of NF-κB by phosphorylation of its inhibitor IκB. Nature 1990;344:678–682.

125 Nabel GJ, Baltimore D: An inducible transcription factor activates expression of human immunodeficiency virus in T cells. Nature 1987;326:711–713.

126 Baldwin AS, Sharp PA: Two transcription factors, NF-κB and H2TF1, interact with a single regulatory sequence in the class I major histocompatibility complex promoter. Proc Natl Acad Sci USA 1988;85:723–727.

127 Singh H, LeBowitz JH, Baldwin AS, Sharp PA: Molecular cloning of an enhancer binding protein: Isolation by screening of an expression library with a recognition site DNA. Cell 1988;52:415–423.

128 Capecchi MR: Altering the genome by homologous recombination. Science 1989; 244:1288–1292.

129 Wabl M, Burrows PD: Expression of immunoglobulin heavy chain at a high level in the absence of a proposed immunoglobulin enhancer element in cis. Proc Natl Acad Sci USA 1984;81:2452–2455.

130 Zaller DM, Eckhardt L: Deletion of a B cell specific enhancer affects transfected, but not endogenous, immunoglobulin heavy chain expression. Proc Natl Acad Sci USA 1985;82:5088–5092.

131 Grosschedl R, Marx M: Stable propagation of the active transcriptional state of an immunoglobulin μ gene requires continuous enhancer function. Cell 1988;55:645–654.

132 Gregor P, Morrison SL: Myeloma mutant with a novel 3′ flanking region: Loss of normal sequence and insertion of repetitive elements lead to decreased transcription but normal processing of the alpha heavy chain gene products. Mol Cell Biol 1986; 6:1903–1916.

133 Stavnezer-Nordgren J, Sirlin S: Specificity of immunoglobulin heavy chain switch correlates with activity of germline heavy chain genes prior to switching. EMBO J 1986;5:95–102.

Patrick Matthias, PhD, Whitehead Institute for Biomedical Research, Nine Cambridge Center, Cambridge, MA 02142 (USA)

Sorg C (ed): Molecular Biology of B Cell Developments.
Cytokines. Basel, Karger, 1990, vol 3, pp 109–125

Induction of Directed Switch Recombination by Cytokines

H. Illges, W. Müller, A. Radbruch[1]

Institute for Genetics, University of Cologne, FRG

Upon activation by mitogen or antigen, 'naive' B lymphocytes expressing IgM and IgD may switch to the expression of immunoglobulins of the IgG, IgA or IgE class. This immunoglobulin class switch changes the effector function of the secreted antibodies, thus increasing the functional diversity of the immune response. Recently, some progress has been made in understanding the regulation and molecular basis of immunoglobulin class switching, providing us with a model system to analyze the regulation of site-directed recombination in lymphocyte differentiation.

This review will focus on recent results and current ideas about the control of immunoglobulin class switching by T-cell-derived cytokines. It is known for long that, in the mouse, T-cell-dependent humoral immune responses are dominated by IgG1 or IgG2a, and that T cells are also involved in the regulation of the production of IgA and IgE, while T-cell-independent responses are dominated by IgG3 and IgG2b [1–4]. How do T cells exert their regulatory function? Today this question can, at least in part, be answered due to the identification and characterization of several T-cell-derived cytokines. Cytokines effective in the regulation of immunoglobulin class switching have been identified and the molecular mechanism of the regulation becomes gradually elucidated.

[1] We thank U. Ringeisen for the graphs and B. Hampel and S. Irlenbusch for expert technical help. Thanks to C. Esser, S. Jung, G. Siebenkötten and J. Schmitz for discussions. This work was supported by the 'Bundesministerium für Forschung und Technologie'. A.R. holds a 'Forschungs-Dozentur' of the Bayer AG.

Cytokines Induce and Suppress Immunoglobulin Class Switching

Due to difficulties in the analysis of cytokine action in vivo, most of our present knowledge about the regulatory effects of cytokines comes from in vitro studies, mainly of murine B cells which can be polyclonally activated to switch in vitro.

On a cellular basis, the regulation by cytokines could either involve a selection of switched cells to proliferate and/or secrete immunoglobulin, or involve a direct induction of the class switch. As expected for pleiotropic cytokines, both effects are observed. Many earlier studies, quantitating secreted immunoglobulin in the culture supernatant or quantitating immunoglobulin mRNA of cultured cells to evaluate the effect of cytokines, could not really separate 'selection' from 'induction' [5–8]. Several groups have used limiting-dilution techniques to determine the frequency of cells that are induced to class switching, and their clonal expansion in cultures of murine B cells polyclonally activated with bacterial lipopolysaccharide (LPS) and cultured in the presence or absence of various cytokines [9–12]. Thus, the induction of cells to switch could be separated from stimulation of secretion of immunoglobulin, although it should be kept in mind that sometimes an apparent increase in frequency might be due to stimulation of secretion because then wells with less secreting cells pass the limit of detection.

The limiting-dilution analyses shows that, first of all, the majority of switched cells in LPS cultures are not derived from the few preexisting IgG-, IgA- or IgE-expressing cells, which are mostly not responsive to LPS. Instead, the switched cells are derived from IgM- and IgD-expressing naive B cells which respond to the mitogen LPS. Switched cells can be detected in the clonal progeny of more than 80% of the activated cells. In some clones, cells expressing various immunoglobulin classes other than IgM are generated.

In the absence of T-cell-derived cytokines, the activated B cells will switch to IgG3 and IgG2b at high frequency, e.g. 30% to IgG3 cells at day 7 of culture, but less to other classes [13, 14]. In the presence of murine interleukin-4 (IL-4), the dominant class is IgG1 instead of IgG3 [11]. IgE-producing cells are also generated at high concentrations of IL-4 [15–17]. T-cell-derived interferon-γ suppresses the effects of IL-4 and may induce IgG2a [8]. IL-4 and IL-5 probably also stimulate the secretion of IgG1, IgE and IgA, respectively [18]. Recently, T cell growth factor-β has been identified as an IgA-switch-inducing factor [19]. As regards IL-4, it has been shown that the

Table 1. Cell cycle kinetics of IgG1-expressing cells

Cell population	Percent cells in S/G$_2$ phase	
	– Colcemid	+ Colcemid
Total LPS blasts	16	24
Cχ^+ LPS blasts	16	24
Cγ_1^+ LPS blasts	16	24

Murine spleen cells were cultivated in the presence of IL-4-containing supernatant and LPS for 5 days. To half of the culture Colcemid was added during the last 8 h of culture. The cells were harvested, fixed in 70% methanol containing 10 µg/ml Hoechst 33342 and subsequently stained cytoplasmically for either χ or γ_1 antibodies. The stained cell population was analyzed on a modified FACS 440 using a dual laser system (351 and 488 nm) by double fluorescence. The number of cells present in the S and G$_2$ phase of the cell cycle for the various cell populations analyzed was determined by the software program build into the FACS 440.

cytokine has to be added within 2 days before or after the activation of B cells by mitogen in order to exert the regulatory effect [16, 20].

Analyzing the cell cycle kinetics of cells expressing IgG1 in the cytoplasm, we could show that, in cultures of B cells stimulated with LPS and IL-4, IgG1 cells progress through the cell cycle at the same average rate as the total population or cells expressing κ-light chains in the cytoplasm, i.e. most B cells (table 1). In addition, IgM and IgG1 cells of a 5-day-old LPS plus IL-4 culture, sorted according to surface immunoglobulin, show the same survival rate when re-cultivated in the presence of IL-4 for another 2 days [A.R., unpubl. results]. Thus, there is no indication that the proliferation or survival of IgG1-expressing cells is enhanced by IL-4 [21].

In summary, apart from stimulating the secretion of IgG, IgA or IgE, cytokines can also induce class switching at the cellular level, as has been shown most clearly for IL-4. Still, at the cellular level, it is not clear whether this induction is due to selection of cells committed to switch to certain classes or whether the cytokine turns uncommitted cells into committed cells. Since not all naive B cells are activated by LPS, the existence of precommitted subpopulations that require a certain cytokine for activation is hard to exclude. At least for IL-4, however, it is clear that all naive B cells have receptors for IL-4. They can be stained with affinity-purified IL-4, as

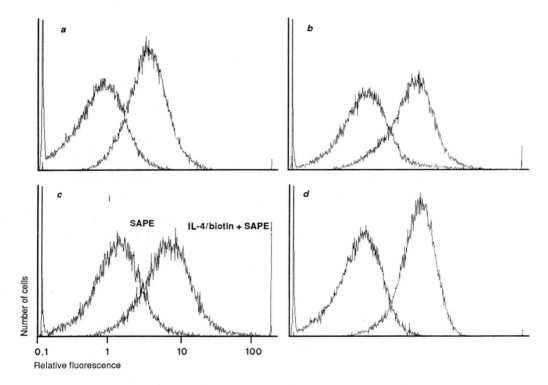

Relative fluorescence

Number of cells

Fig. 1. Flow-cytometric analysis (FACS 440, electric desk) of IL-4 receptor expression on lymphocytes from various organs. Displayed is the number of cells versus the relative fluorescence. Lymphocytes isolated from thymus *(a)*, spleen *(b)*, peritoneal cavity *(c)* and mesenteric lymph nodes *(d)* were either stained with streptavidin-phycoerythrin (SAPE) or first with recombinant, affinity-purified, biotinylated murine IL-4 and subsequently with SAPE. In the organs analyzed, all cells express IL-4 receptors. The positive population is homogeneous. A cell population negative for IL-4 receptor expression is not detectable. The staining is specifically inhibited by IL-4, not by other cytokines.

shown in figure 1. Moreover, all naive B cells respond to stimulation with IL-4 by increased expression of MHC class II antigen within 16 h [22]. Thus, it seems likely that IL-4, and other cytokines alike, act on all B cells and commit them to switch to distinct classes of immunoglobulin upon activation by antigen or mitogen. This view is suported by molecular analyses of immunoglobulin class switching.

Molecular Basis of Immunoglobulin Class Switching

Immunoglobulin class switch recombination is the deletion of heavy-chain constant region (C_H) genes by recombination of specialized DNA sequences, the switch (S_H) regions, which are located before (5') all C_H genes, except $C\delta$ of the mouse [23]. Activated and switched B cells have relocated the $C\gamma$, $C\alpha$ or $C\varepsilon$ genes to be expressed next to the VDJ gene by switch recombination, and deleted the $C\mu$ and other intermittent C_H genes. Class switching without switch recombination, but by differential transcription, has been reported. The evidence for this to occur is not clear yet, as we have discussed elsewhere [24].

Directed Class Switch Recombination

Class switch recombination is dependent on cell division [25] and thus will occur only in B cells activated by mitogen or antigen for proliferation. The regulation of class switching by cytokines has to effect switch recombination, although no T-cell-derived cytokine has been described which can activate B cells for proliferation and switch recombination. It is not switch recombination as such which is controlled by cytokines, but the kind of switch regions that participate in the recombination. This has been analyzed first for murine B cells activated with LPS and induced to switch to IgG1 by IL-4. Comparing switch recombinations on active and inactive immunoglobulin heavy-chain *(Igh)* loci of such cells isolated by cell sorting, it turned out that on both *Igh* loci switch recombination had occurred mainly between the $S\mu$ and $S\gamma_1$ switch regions [26, 27]. Thus, it became clear that IL-4 directs switch recombination to the $S\gamma_1$ region. Since then, analyses of B cells activated with LPS alone [27–30] and of B cells immortalized by cell fusion, after class switching in various in vivo and in vitro systems [26, 27, 31, 32], have confirmed that class switch recombination on the inactive *Igh* locus is directed to the same switch region as on the active locus in at least 60–80% of the cells that show switch recombination on the inactive locus at all.

Switch recombination could be directed to discrete switch regions either by induction of specific recombinases, one for each switch program, or by targeting of a 'universal' switch recombinase to distinct switch regions. Several groups have put considerable effort in the analysis of possible sequence requirements for switch recombination.

Switch regions are composed of repeated short sequences, variations on motifs like AGCT, TAGAGC, GAGCT, TGGGG, GGGGT, GAGCTGGGG, ACCAG and GCAGC [23, 33–48]. Sμ has strong homology to Sα and Sε, and less to the Sγ regions which have considerable homology among themselves. They consist of repeats of a characteristic 49-basepair (bp) sequence, which is repeated from 44 times (Sγ$_3$ [33, 45]) to more than 120 times (Sγ$_1$ [34]). In addition to the 49-bp repeats, the Sγ$_1$ region contains two direct and two long identical repeats [34]. This characteristic organization of the Sγ$_1$ region is conserved for all *Igh* haplotypes, which otherwise show extensive length polymorphism [34]. The functional significance of this conservation is not clear. All attempts to identify specific signal sequences for switch recombination, which should be located close to the recombination sites, until now showed no conclusive result [29, 30, 49–51], especially when comparing Sμ/Sγ$_3$ and Sμ/Sγ$_1$ recombinations, the latter being induced by IL-4 [27, 33]. Instead, the analysis shows that the recombination can take place throughout the switch region and for Sμ is sometimes also found outside the repetitive sequences [27].

The lack of apparent signal sequences for switch-region-specific switch recombinases suggests that there might be only one switch recombinase, which is somehow directed to distinct switch regions. How could switch regions be targeted for switch recombination? It is known from Rpo (RNA-polymerase)-mediated recombination in phage λ [52], the mating-type switch in yeast [53] and studies on the expression of unrearranged V genes [54] that transcription of recombinogenic sequences is correlated with recombination events. The conformational changes of DNA induced by transcription [55–57], as well as experimental systems in which transcribed DNA, dependent on the activity of topoisomerases I or II, is recombined at high frequency [58, 59], argue for the hypothesis that the site-specific induction of torsional stress in transcribed DNA may generate recombinogenic structures.

Cytokines Regulate Transcription of Switch Regions

Several groups have reported evidence that immunoglobulin switch regions are transcribed before switch recombination [60–67]. Transcription starts 5′ of the respective switch region and procedes through the switch region and the C$_H$ gene. The transcripts are processed and transported to the cytoplasm. Such transcripts have been first detected in the murine cell lines 18–81 and I29 which show spontaneous class switching preferentially to

IgG2b and IgA, respectively. In 18–81, transcripts of the $S\gamma_{2b}$ and in I29, transcripts of the $S\alpha$ region but less or none of the other switch regions are detectable [60, 61, 63]. In naive splenic B cells and B cells activated with LPS, only the $S\gamma_3$ and $S\gamma_{2b}$ regions are transcribed [64, 65]. In B cells and LPS blasts stimulated with IL-4, transcription of $S\gamma_1$ and, at higher doses of IL-4, also of $S\epsilon$ is induced [62, 67]. Interferon-γ can inhibit the effect of IL-4 [65]. In summary, the appearance of transcripts of distinct switch regions correlates with class switch recombination of these switch regions and is induced or suppressed by cytokines.

The analysis of proteins and DNA sequences involved in the regulation of transcription of switch regions is just in its infancy. As far as analyzed, the transcripts start at multiple sites upstream of the switch regions [63, 66, 68]. At least for $S\gamma_1$, $S\gamma_{2b}$, $S\alpha$ and $S\epsilon$, no 'classic' promoter or enhancer sequence motifs have been identified unambiguously, although sequences reminiscent of some enhancer motifs have been described [68]. An IL-4-inducible DNase-I-hypersensitive site is located at the site of initiation of switch region transcription [69]. Upstream of that site, a nuclear protein binds to the DNA, induced by IL-4 [H.I. and A.R., in preparation]. Another DNA-binding protein in the vicinity of the $S\alpha$ region has been recently described by the group of Stavnezer [70].

The regulation of class switch recombination may also involve demethylation of switch region DNA. Many studies have shown that transcriptional control regions of genes tend to be undermethylated in cells in which the genes are expressed relative to cells in which they are not expressed [71–76]. Analyses of methylation of switch region DNA in the B cell line I29, which may switch spontaneously from IgM to IgA, showed that $S\gamma_{2a}$, $S\epsilon$ and SA, but not $S\gamma_{2b}$ or $S\gamma_1$, are demethylated in I29 cells compared to liver cells [60]. The methylation of the $C\gamma_1$ gene and $S\gamma_1$ switch region in T and B lymphocytes and in nonlymphoid cells has been studied by Burger and Radbruch [submitted]. Stimulation of LPS-activated B cells with IL-4 results in specific demethylation of a stretch of DNA upstream of $S\gamma_1$, but not of $C\gamma_1$. The demethylated DNA coincides with the region of initiation of transcription [Burger and Radbruch, submitted; 66, 69].

In T cells, B cells and B cells activated with LPS alone, the DNA of the $S\gamma_1$ and $C\gamma_1$ region is methylated to an even higher degree than in nonlymphoid cells. This could be interpreted as 'protective' methylation, preventing accidental switch recombination in the lymphoid cell lineage [Burger and Radbruch, submitted].

Structure and Function of Switch Region Transcripts

Several transcripts resulting from transcription of switch regions have been cloned and characterized [63, 68, 77, 87]. Comparing such transcripts from several murine and human switch regions, some basic structural features become obvious. The transcripts start at multiple sites upstream of the switch regions (see above) and extend through the corresponding C_H gene, terminating at the polyadenylation signals downstream of the membrane and secretion exons. The transcripts are processed to give an mRNA with a pseudo (ψ) exon, encoded upstream of the switch region, spliced to the exons of the C_H gene in its membrane and secreted form. It has been postulated that such transcripts might be used for transsplicing, connecting the VDJ exon of the VDJ $C\mu$ transcript to the $C\gamma$, $C\alpha$ for $C\varepsilon$ exons of the switch region transcripts [68], resulting in a 'switch without switch recombination'.

A comparison of the sequence of the ψ-exon before $S\gamma_{2b}$, also termed $I\gamma_{2b}$ [63], and of the corresponding sequence upstream of $S\gamma_1$ [34] reveals a strong homology of more than 70% (fig. 2). Two human switch region ψ-exons, $I\gamma_1$ and $I\gamma_3$, show a similar homology among themselves and, moreover, parts of their sequences are homologous to parts of the murine $I\gamma_{2b}$ sequence [68]. The striking conservation of parts of the intron sequences and their similar localization [62–68] point to an important function which is under strong selective pressure.

Since the ψ-exons contain multiple stop codons and no long open reading frames, if any, only short peptides could be generated from them [63, 68, 77]. As yet, no such peptide could be identified and shown to play a role in switch recombination. Another reason for conservation of $I\gamma$ DNA sequences could be that they might contain signals for transcription, i.e. enhancer/promoter sequences. This has been demonstrated for a similar transcript which is produced from the *Igh* enhancer region early in B cell ontogeny and contains a ψ-exon, located upstream of $S\mu$ and the $C\mu$ exons ($I\mu$ [78]). Sideras et al. [68] have described enhancer core-like sequences in the human $I\gamma_1$ and $I\gamma_3$ transcripts. Enhancer elements conferring the cytokine responsiveness should be specific for individual switch regions and not be shared by all of them. However, they could well be located close to the 'public' sequences and work in collaboration with them. A third possible function for switch region transcription has been recently discussed [79–81], suggesting that switch region RNA might be involved in the formation of secondary structures of switch region DNA, this facilitating recombination. Theoretically, DNA of switch regions could form triple-stranded structures. These triple strands

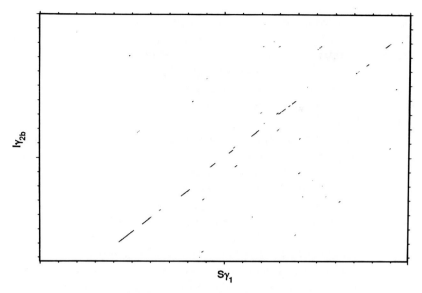

Fig. 2. Comparison of the nucleotide sequences of the $I\gamma_{2b}$ ψ-exon [63] with the sequence of $S\gamma_1$ [34]. Sequence comparison was done by the use of the compare and dotplot program [90]. Parameters: window = 21; stringency = 14.0.

could be composed just of DNA [82], or of DNA and RNA strands. They are stable enough to confer SI hypersensitivity to the participating sequences. Collier et al. [80] demonstrated that part of the murine $S\alpha$ switch region, when cloned into a recombinant plasmid, showed a nuclease SI hypersensitive site within the direct repeat II, concluding that intramolecular triple strands had been formed upon transcription. Later, Griffin and Reabau [pers. commun.] could show that the proposed triple-strand structure can be stabilized by homologous RNA.

Molecular Mechanism of Class Switch Recombination

Pending the isolation of a switch recombinase and elucidation of the role of switch region transcription for switch recombination, today the mechanism of switch recombination can be described only in general terms. As has been mentioned above, no discrete sequence requirements for switch recombination have been unequivocally identified. Recombination breakpoints are found all over the switch regions, at a variety of sequences. No apparent characteristic sequence motif is associated with the breakpoints. Nevertheless, the repetitive structure of the switch regions might be an important factor in facilitating recombination, as is suggested by transfection studies [83]. Switch region DNA will easily form unusual secondary structures that might enhance recombination, like single strands in B-Z DNA (right-handed/left-handed DNA double helix) junctions, and cruciform or slipped DNA stuctures [79].

Since for recombination the switch regions have to be allocated over large distances, not only switch regions from the same chromatid, i.e. *Igh* locus, but also from the *Igh* locus of the sister chromatid or the other chromosome could be involved in recombination, as illustrated in figure 3. Recombination within the same *Igh* locus seems to be the most frequent event, the intervening DNA being deleted (fig. 3a) [27]. Inversion instead of deletion has been described for switching cells from the murine cell line 18–81 [50, 84], but seems to be the exception rather than the rule. In normal B cells, inversion instead of deletion would in all likelihood silence the affected *Igh* locus. Switch recombination by unequal sister chromatid exchange (fig. 3b) may occur occasionally [51], but is hard to demonstrate because the 'switched' chromatid in the switched cell is indistinguishable from one that is generated by deletion, while the sister chromatid with part of the *Igh* locus duplicated is distributed to the IgM-expressing sister cell, and this cell will be identified and analyzed only by chance. Transchromosomal recombination is easier to access if both *Igh* haplotypes show serological and/or restriction polymorphism. Cells that underwent transchromosomal switch recombination have been identified as switch variants, occurring spontaneously in hybridoma cell lines [85, 86]. Transchromosomal recombination is also responsible for translocation of the c-*myc* gene to switch region sequences, frequently the same switch region as is involved in recombination on the other *Igh* locus of the same cell. This is consistent with the idea that c-*myc* translocation can occur as an accident of normal, directed switch recombination [21, 27, 88]. While the frequency of transchromosomal switch recombination in hybridomas is rather low, frequent transchromosomal switching is described for transgenic mice with a VDJ, $S\mu$ and $C\mu$ transgene, integrated outside of the *Igh* locus [89]. $S\mu$ of the transgene and $S\gamma_1$ of the autologous *Igh* locus underwent switch recombination. It remains to be shown whether the high frequency of transchromosomal recombination in this system may be due to selective pressure. In normal murine or human B cells, transchromosomal recombination still has to be demonstrated.

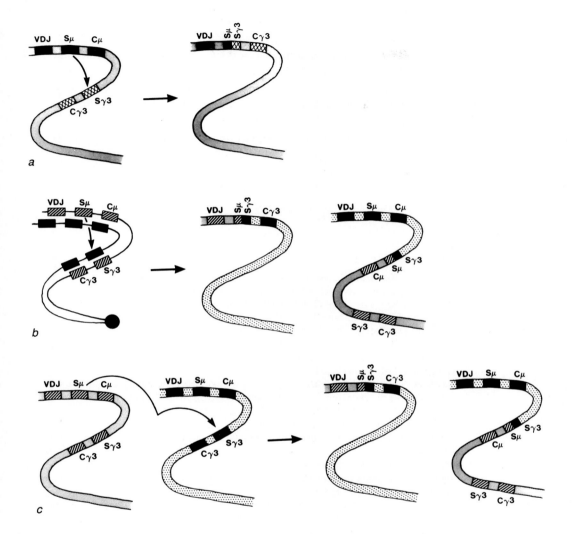

Fig. 3. Genetic classification of various switch recombination mechanisms. *a* Deletion of C_H genes from the same chromatid by the loop out mechanism [50] is probably the most common event. *b* By unequal sister chromatid exchange, C_H genes are deleted from one chromatid and duplicated on the sister chromatid. The two chromatids are separated in mitosis and distributed to the daughter cells. *c* Transchromosomal recombination will join C_H genes from the parental chromosome with the inactive *Igh* locus (or c-*myc* locus) to the VDJ gene of the chromosome with the active *Igh* locus of a cell.

Perspectives

Future analyses of immunoglobulin class switching, aiming at molecular dissection of the process of switch recombination, will have to show how transcription of switch regions, specifically induced by cytokines, is linked to the recombination event. Switch recombination, thus, may exemplify the recombination of immunoglobulin and T cell receptor gene segments in lymphocyte ontogeny. In all likelihood, not only cytokines can act as signals for induction of recombination, although today attention is focused on soluble mediators. Cell membrane-bound 'cytokines' yet to be discovered, may be as potent for the control of recombination.

References

1   Cebra JJ, Komisar JL, Schweitzer PA: C_H isotype switching during normal B-lymphocyte development. Annu Rev Immunol 1984;2:493–548.
2   Perlmutter RM, Hansburg D, Briles DE, Nicolotti RA, Davie JM: Subclass restriction of murine anti-carbohydrate antibodies. J Immunol 1978;121:566–572.
3   Slack J, Der-Balian GP, Nahm M, Davie JM: Subclass restriction of murine antibodies: II. The IgG plaque-forming cell response to thymus-independent type 1 and type 2 antigens in normal mice and mice expressing an X-linked immunodeficiency. J. Exp Med 1980;151:853–862.
4   Tesch H, Smith FI, Müller-Hermes WJP, Rajewsky K: Heterogeneous and monoclonal helper T cells induce similar anti-(4-hydroxy-3-nitro-phenyl)acetyl (NP) response: I. Isotype distribution. Eur J Immunol 1984;14:188–194.
5   Isakson PC, Pure E, Vitetta ES, Krammer PH: T-cell-derived B cell differentiation factor(s) (BCDF): Effect on the isotype switch of murine B cells. J Exp Med 1982; 155:734–748.
6   Müller W, Kühn R, Goldmann W, Tesch H, Smith FI, Radbruch A, Rajewsky K: Signal requirements for growth and differentiation of activated murine B lymphocytes. J Immunol 1985;135:1213–1219.
7   Yuan DE, Weiss A, Layton JE, Krammer PH, Vitetta ES: Activation of γ1 gene by lipopolysaccharide and T cell-derived lymphokines containing a B cell differentiation factor for IgG1 (BCDFγ). J Immunol 1985;135:1465–1472.
8   Snapper CM, Paul WE: Interferon γ and B cell stimulatory factor-1 reciprocally regulate Ig isotype production. Science 1987;236:944–947.
9   Anderson J, Coutinho A, Melchers F: The switch from IgM to IgG secretion in single mitogen-stimulated B-cell clones. J Exp Med 1978;147:1744–1754.
10 Coutinho A, Forni L: Intraclonal diversification in immunoglobulin isotype secretion: An analysis of switch probabilities. EMBO J 1982;1:1251–1257.
11 Layton JE, Vitetta ES, Uhr JW, Krammer PH: Clonal analysis of B cells induced to secrete IgG by T-cell derived lymphokine(s). J Exp Med 1984;160:1850–1863.
12 Bergstedt-Lindqvist S, Moon H-B, Persson U, Möller G, Heusser C, Severinson E:

Interleukin 4 instructs uncommitted B lymphocytes to switch to IgG1 and IgE. Eur J Immunol 1988;18:1073–1077.

13 McHeyzer-Williams HG: Combinations of interleukins 2, 4 and 5 regulate the secretion of murine immunoglobulin isotypes. Eur J Immunol 1989;19:2025–2030.

14 Kearney JF, Cooper MD, Lawton AR: B cell differentiation induced by lipopolysaccharide: IV. Development of immunoglobulin class restriction in precursors of IgG-synthesizing cells. J Immunol 1976;117:1567–1572.

15 Radbruch A, Sablitzky F: Deletion of Cμ genes in mouse B lymphocytes upon stimulation with LPS. EMBO J 1983;2:1929–1935.

16 Savellkoul HJF, Lebman DA, Benner R, Coffmann RL: Increase of precursor frequency and clonal size of murine IgE-secreting cells by IL-4. J Immunol 1988;141: 749–755.

17 Snapper CM, Finkelman FD, Paul WE: Differential regulation of IgG1 and IgE synthesis by interleukin 4. J Exp Med 1988;167:183–196.

18 Coffman RL, Mosmann TR: Isotype regulation by helper T cells and lymphokines; in Dukor P, Hanson LA, Kallós P, Shakilo F, Trnka Z, Walksman BH (eds): Monogr Allergy. Basel, Karger, 1988, vol 24, pp 96–103.

19 Sonoda E, Matsumoto R, Hitoschi Y, Tishii T, Sugimoto M, Araki S, Tominagawa A, Yamaguichi N, Tanatsu K: Transforming growth factor β induces IgA production and acts additively with interleukin 5 for IgA production. J Exp Med 1989;170: 1415–1420.

20 Snapper CM, Paul WE: B cell stimulatory factor-1 (interleukin 4) prepares resting murine B cells to secrete IgG1 upon subsequent stimulation with bacterial lipopolysaccharide. J Immunol 1987;139:10–17.

21 Radbruch A, Burger C, Klein S, Müller W: Control of immunoglobulin class switch recombination. Immunol Rev 1986;89:69–84.

22 Noelle RJ, Kuziel WA, Maliszewski CR, McAdams E, Vitetta ES, Tucker PW: Regulation of the expression of multiple class II genes in murine B cells by B cell stimulatory factor-1 (BSF-1). J Immunol 1986;137:1718–1723.

23 Shimizu A, Takahashi N, Yaoita Y, Honjo T: Organisation of the constant region gene family of the mouse immunoglobulin chain. Cell 1982;28:499–506.

24 Esser C, Radbruch A: Immunoglobulin class switch: Molecular and cellular analysis. Annu Rev Immunol 1990;8:717–735.

25 Severinson-Gronowicz E, Doss C, Schröder J: Activation to IgG secretion by lipopolysaccharide requires several proliferative cycles J Immunol 1979;123:2057–2067.

26 Radbruch A, Müller W, Rajewsky K: Class switch recombination is IgG1 specific on active and inactive IgH loci of IgG1 secreting B cell blasts. Proc Natl Acad Sci USA 1986;83:3954–3957.

27 Winter E, Krawinkel K, Radbruch A: Directed Ig class switch recombination in activated murine B cells. EMBO J 1987;6:1663–1671.

28 Kepron MR, Chen YW, Uhr JW, Vitetta ES: IL4 induces the specific rearrangement of X1 genes on the expressed and unexpressed chromosomes of lipopolysaccharide-activated normal murine B cells. J Immunol 1989;143:334–339.

29 Radbruch A, Sablitzky F: Deletion of Cμ genes in mouse B lymphocytes upon stimulation with LPS. EMBO J 1983;2:1929–1935.

30 Hurwitz JL, Cebra JJ: Rearrangements between the immunoglobulin heavy chain gene J and Cμ regions accompany normal B lymphocyte differentiation in vitro. Nature 1982;299:742–744.

31 Schultz C, Petrini J, Collins J, Claflin JL, Denis KA, Gearhart P, Gritzmacher C, Manser T, Shulman M, Dunnick W: Patterns and extent of isotype specificity in the murine heavy chain switch DNA rearrangement. J Immunol, in press.

32 Hummel M, Berry JK, Dunnick W: Switch region content of hybridomas: The two spleen cell Igh loci tend to rearrange to the same isotype. J Immunol 1987;138:3539–3548.

33 Szurek P, Petrini TJ, Dunnick W: Complete nucleotide sequence of the murine γ3 switch region and analysis of switch recombination sites in two γ3 expressing hybridomas. J Immunol 1985;135:620–626.

34 Mowatt MR, Dunnick WA: DNA sequence of the murine γ1 switch segment reveals novel structural elements. J Immunol 1986;136:2674–2683.

35 Kataoka T, Miyata T, Honjo T: Repetitive sequences in class-switch recombination regions of immunoglobulin heavy chain genes. Cell 1981;23:357–358.

36 Lang RB, Stanton LW, Marcu KB: On immunoglobulin heavy chain gene switching: Two γ2b genes are rearranged via switch sequences in MPC11 cells but only one is expressed. Nucleic Acids Res 1982;10:611–630.

37 Sakano H, Maki R, Kurosawa Y, Roeder W, Tonegawa S: Two types of somatic recombination are necessary for the generation of complete immunoglobulin heavy chain genes. Nature 1980;286:676–683.

38 Davis MM, Kim SK, Hood LE: DNA sequences mediating class switching in α immunoglobulins. Science 1980;209:1360–1368.

39 Cory S, Adams JM: Deletions are associated with somatic rearrangements of immunoglobulin heavy chain genes. Cell 1980;19:37–51.

40 Nikaido T, Nikai S, Honjo T: Switch region of immunoglobulin Cμ gene is composed of simple tandem repetitive sequences. Nature 1981;292:845–848.

41 Gough NM, Bernard O: Sequences of the joining region genes for immunoglobulin heavy chains and their role in generation of antibody diversity. Proc Natl Acad Sci USA 1981;78:509–513.

42 Obata M, Kataoka T, Nakai S, Yamagishi H, Takahashi N, Yamawaki-Kataoka Y, Nikaido T, Shimizu A, Honjo T: Structure of a rearranged γ1 chain gene and its implication to immunoglobulin class switch mechanism. Proc Natl Acad Sci USA 1981;78:2437–2441.

43 Gillies SD, Morrison SL, Oi VT, Tonnegawa S: A tissue specific transcription enhancer element is located in the major intron of a rearranged immunoglobulin heavy chain gene. Cell 1983;33:717–728.

44 Greenberg R, Lang RB, Diamond MS, Marcu KB: A switch region inversion contributes to the aberrant rearrangement of a IgM immunoglobulin heavy chain gene in MPC-11 cells. Nucleic Acids Res 1982;10:7751–7761.

45 Stanton LW, Marcu KB: Nucleotid sequence and properties of the murine γ3 immunoglobulin heavy chain gene switch region: Implications for successive Cγ gene switching. Nucleic Acids Res 1982;10:5993–6006.

46 Kataoka T, Kawakami T, Takahashi N, Honjo T: Rearrangement of immunoglobulin γ1-chain gene and mechanism for heavy-chain class switch. Proc Natl Acad Sci USA 1980;77:919–923.

47 Nikaido T, Yamawaki-Kataoka Y, Honjo T: Nucleotide sequences of switch regions of immunoglobulin Cε and Cγ genes and their comparison. J Biol Chem 1982; 257:7322–7329.

48 Takashi N, Kataoka T, Honjo T: Nucleotide sequences of class-switch recom-

bination region of the mouse immunoglobulin γ2b-chain gene. Gene 1980;11:117–121.

49   Petrini J, Dunnich WA: Products and implied mechanism of H chain switch recombination. J Immunol 1989;142:2932–2935.

50   Jäck HM, McDowell M, Steinberg CM, Wabl M: Looping out and deletion mechanism for the immunoglobulin heavy chain class switch. J Immunol 1988;85:1581–1585.

51   Katzenberg DR, Shermaine AT, Birshstein BK: Nucleotide sequence of an unequal sister chromatid exchange site in mouse myeloma cell line. Mol Cell Biol 1989;9:1324–1326.

52   Ikeda H, Matsumato T: Transcription promotes rec A independent recombination mediated by DNA dependent RNA polymerase in *Escherichia coli*. Proc Natl Acad Sci USA 1979;76:4571–4575.

53   Voelkel-Meiman K, Keil ML, Roeder GS: Recombination-stimulating sequences in yeast ribosomal DNA correspond to sequences regulating transcription by RNA polymerase I. Cell 1987;48:1071–1079.

54   Yancopoulos GD, Alt FW: Developmentally controlled and tissue specific expression of unrearranged V gene segments. Cell 1985;40:271–281.

55   Wells RD, Goodman TC, Hillen W, Horn GT, Klein RD, Larson JE, Müller UR, Neuendorf SK, Panyotos N, Stirdivants M: DNA structure and gene regulation. Prog Nucleic Acids Res 1980;24:167–267.

56   Giaever GN, Wang JC: Supercoiling of intracellular DNA can occur in eukaryotic cells. Cell 1989;55:849–856.

57   Tsao Y-P, Wu H-J, Liu LF: Transcription-driven supercoiling of DNA: Direct biochemical evidence from in vitro studies. Cell 1989;56:111–118.

58   Kim RA, Wang JC: A subthreshold level of DNA topoisomerases leads to the excision of yeast rDNA as extrachromosomal rings. Cell 1989;57:975–985.

59   Fink GR: A new twist to the topoisomerase I problem. Cell 1989;58:25–26.

60   Stavnezer-Nordgren J, Sirlin S: Specificity of immunoglobulin heavy chain switch correlates with activity of germline heavy chain genes prior to switching. EMBO J 1986;5:95–102.

61   Alt FW, Blackwell TK, DePinho RA, Reth MG, Yancopoulos GD: Regulation of genome rearrangement events during lymphocyte differentiation. Immunol Rev 1986;89:5–30.

62   Rothman P, Lutzker S, Cook W, Coffman R, Alt FW: Mitogen plus interleukin 4 induction of Cε transcripts in B lymphoid cells. J Exp Med 1988;168:2385–2389.

63   Lutzker S, Alt FW: Structure and expression of germline immunoglobulin γ2b transcripts. Mol Cell Biol 1988;8:1849–1852.

64   Lutzker S, Rothman P, Polloch R, Coffman R, Alt FW: Mitogen and IL-4 regulated expression of germline γ 2b transcripts: Evidence for directed heavy chain class switching. Cell 1988;53:117–184.

65   Esser C, Radbruch A: Rapid induction of transcription of unrearranged sγ1 switch regions in activated murine B-cells by interleukin 4. EMBO J 1989;8:483–488.

66   Berton MT, Uhr JW, Vitetta ES: Synthesis of germline γ1 immunoglobulin heavy chain transcripts in resting B-cells: Induction by interleukin 4 and inhibition by interferon γ. Proc Natl Acad Sci USA 1989;86:2829–2833.

67 Rothman P, Lutzker S, Li S, Coffman R, Alt FW: Mitogen plus IL-4 regulation of heavy chain germline transcripts: Evidence for directed class-switching (abstract 63–50). 7th Int Congr Immunol 1989, Stuttgart, Fischer, 1989.

68 Sideras P, Mizuta T-R, Kanamori H, Suzuki N, Okamoto M, Kuze K, Ohno H, Doi S, Fukuhara S, Hassan MS, Hammarström L, Smith E, Shimizu A, Honjo T: Production of sterile transcripts of cγ gene in an IgM-producing B cell line that switches to IgG-producing cells. Int Immunol 1989;1:631–641.

69 Schmitz J, Radbruch A: An interleukin 4 induced DNAse I hypersensitive site indicates opening of the γ1 switch region prior to switch recombination. Int Immunol 1989;1:570–575.

70 Waters HS, Sail UK, Stavnezer J: A B cell specific nuclear protein binds to DNA sites 5′ to immunoglobulin Sα tandem repeats is regulated during differentiation. Mol Cell Biol 1989;9:5594–5601.

71 Cedar H: DNA methylation and gene activity. Cell 1988;53:3–4.

72 Dynan WS, Tjian R: Control of eukaryotic messenger RNA synthesis by sequence-specific DNA binding proteins. Nature 1985;316:774–778.

73 Holler M, Westin G, Jiricny J, Schaffner W: Sp1 transcription factor binds DNA and activates transcription even when the binding site is CpG methylated. Genes Dev 1988;2:1127–1135.

74 Toth M, Lichtenberg U, Doerfler W: Genomic sequencing reveals a 5-methylcytosine-free domain in active promotors and the spreading of preimposed methylation patterns. Proc Natl Acad Sci USA 1989;86:3728–3732.

75 Dynan WS: Understanding the molecular mechanism by which methylation influences gene expression. TIG 1989;5:35–36.

76 Holliday R: The inheritance of epogenetic defects. Science 1987;238:163–170.

77 Radcliffe G, Lin Y-C, Julius M, Marcu KB, Stavnezer J: Structure of immunoglobulin α heavy chain RNA and its location on polysomes. Mol Cell Biol, in press.

78 Lennon GG, Perry RP: Cμ containing transcripts initiate heterogeneously within the IgH enhancer region and contain a novel 5′ untranslatable exon. Nature 1985; 318:475–476.

79 Wells RD, Anurhaeri S, Blaho JA, Collier DA, Hanvey JC, Hsieh W-T, Jaworski A, Klysik J, Larson JE, McLean JM, Wohlrab F, Zacharias W: Unusual DNA structures and the probes used for their detection; in Wells RD, Harvey SC (eds): Unusual DNA Structures. Berlin, Springer, 1988, pp 1–22.

80 Collier DA, Griffin JA, Wells RD: Non B right handed DNA conformations of homopurine-homopyrimidine sequences in the murine immunoglobulin Cα switch region. J Biol Chem 1987;263:7397–7405.

81 Griffin JA, Reabau ME: A role for transcription in antibody switch recombination; in Gottesman ME, Vogel H (eds): Mechanisms of Eukaryotic DNA recombination. P + S Biomed Symp, 1989, in press.

82 Mirkin SM, Lyamichev VI, Druschlyak KN, Dobrynin VN, Filippov SA, Frank-Kamenetski MD: DNA H form requires a homopurine-homopyrimidine mirror repeat. Nature 1987;330:495–497.

83 Ott DE, Alt FW, Marcu KB: Immunoglobulin heavy chain switch region recombination within a retroviral vector in murine pre-B cells. EMBO J 1987;6:577–584.

84 Burrows PD, Beck GB, Wabl MR: Expression of μ and γ immunoglobulin heavy chains in different cells of a cloned mouse lymphoid line. Proc Natl Acad Sci USA 1981;78:564–568.

85 Sablitzky F, Radbruch A, Rajewsky K: Spontaneous immunoglobulin class switching in myeloma and hybridoma cell lines differs from physiological class switching. Immunol Rev 1982;67:59–72.

86 Kipps TJ, Herzenberg LA: Homologous chromosome recombination generating immunoglobulin allotype and isotype switch variants. EMBO J 1986;5:263–268.

87 Gerondakis S: Structure and expression of murine germline immunoglobulin ε heavy chain transcripts induced by interleukin 4. Proc Natl Acad Sci USA 1990;87:1581–1585.

88 Tian SS, Faust C: Rearrangement of rat immunoglobulin E heavy chain and c-myc genes in the B cell lymphoma IR 162. Mol Cell Biol 1987;7:2614–2619.

89 Durdik J, Gerstein RM, Rath S, Robbins S, Nisonoff A, Selsing E: Isotype switching by a microinjected μ immunoglobulin heavy chain gene in transgenic mice. Proc Natl Acad Sci USA 1989;86:2346–2350.

90 Maizel X, Lenk Y: Enhanced graphic matrix analysis of nucleic acid. Proc Natl Acad Sci USA 1981;78:7665–7669.

Harald Illges, Institut für Genetik, Universität zu Köln,
Weyertal 121, D–5000 Köln 41 (FRG)

Sorg C (ed): Molecular Biology of B Cell Developments.
Cytokines. Basel, Karger, 1990, vol 3, pp 126–131

Differentiation of Pre-B Cells in vitro

Influence of Growth Factors on Pre-B Cells before and after
Heavy-Chain Gene Commitment

Helmut Sauter

Unité de Génétique Somatique, Institut Pasteur, Paris, France

The mature cells of the hemopoietic system develop from common
progenitor cells in a series of differentiation events. This process is characterized by expansion in cell numbers and progressive restriction of the developmental potential of the progenitor cells.

Considerable progress has been made in the last years in the identification and characterization of the intermediate cell types of the lymphoid and
myeloid compartments of the hemopoietic system. In addition, a series of
growth factors has been identified which act on hemopoietic progenitor cells
at various stages of differentiation. It is becoming more and more clear that
many cytokines exert multiple effects on cells belonging to different lineages,
thus forming a complex network of interactions between growth-factor-
producing and growth-factor-responsive cells.

A major gap in our understanding of cytokine action is the question on
the nature of growth factors influencing the early lymphoid development.
This lack of knowledge reflects the fact that only recently did in vitro culture
systems for lymphoid precursors become available, allowing the identification of these cytokines.

Clonable Pre-B Cells

The culture systems which are used to study B cell development in vitro
can be subdivided into two groups: short-term and long-term culture systems. The long-term system developed by Whitelock and Witte will be
described in more detail elsewhere in this volume. Here, I will concentrate on

experiments using a short-term agar culture system originally developed by Paige [1]. The system is a modification of previously described methods for the growth of hemopoietic cells in semisolid media [2–7]. Semisolid cultures in methylcellulose or agar have proved particularly useful in the characterization of intermediate cellular stages of myeloid cell lineages, since in these assays one cell gives rise to a colony consisting of more mature progeny. This allows for establishment of parent-progeny relationships between cells. Moreover, these systems have helped to characterize growth factors involved in the differentiation of myeloid cells at various points of development [for review, see ref. 8]. The first agar culture system for B lineage cells has been described by Metcalf et al. [9–11] and Kincade et al. [12]. In the colony-forming unit-B assay mature surface-immunoglobulin-positive B cells generate colonies consisting of immunoglobulin-secreting plasma cells [13]. The system relies on stimulation of B cells by agar mitogens and lipopolysaccharide [12, 14]. Pre-B cells cloned under these conditions do not differentiate into mature B cell colonies. However, if pre-B cells are cultured on an established layer of adherent cells either derived from bone marrow or fetal liver, maturation of B cell precursors to immunoglobulin-secreting cells can be observed [1]. Thus, these adherent cells provide the growth signals required for the transition from the pre-B to the B cell stage.

Clonable Pre-B Cells are a Heterogeneous Population of Cells before and after Commitment to Particular Immunoglobulin Genes

The first question arising in this context is: what kind of pre-B cells differentiate under these conditions? The genetic stage of single primary pre-B cells cannot be determined directly. The agar culture system offers an approach to answer this question. The analysis of transformed pre-B cell lines representative for particular differentiation stages has shown that rearrangement and expression of immunoglobulin genes is an ordered process: after recombination of heavy-chain gene segments, pre-B cells express μ-heavy chains in the cytoplasm. Thereafter, they start to recombine light-chain genes. If this rearrangement is productive, pre-B cells reach the B cell stage by expressing immunoglobulin molecules on the cell surface [15–20]. Based on the phenomena of allelic and isotype exclusion, i.e. that one B cell will produce only one type of heavy chain and light chain, early and late pre-B cells maturing to immunoglobulin-secreting colonies will generate different types of colonies: a pre-B cell after commitment to heavy and light chain will generate a colony in which one type of heavy chain and one type of light chain is produced. A pre-B cell before light-chain gene commitment, but after commitment to a particular heavy-chain gene, will give rise to daughter cells which rearrange different light-chain genes. Therefore, the colony arising from such a pre-B cell will be homogeneous for heavy chain, but heterogeneous regarding light-chain production. As a third possibility, a totally uncommitted pre-B cell will form a colony in which different heavy and different light chains are expressed at the same time.

In order to perform this analysis, the secreted immunoglobulin from single colonies is transferred to nitrocellulose filters. B cell colonies are then visualized by developing the filters with enzyme-labeled anti-immunoglobulin antibodies. Since the method allows replicate blotting from the same colony, the heterogeneity of the colonies as concerns immunoglobulin expression can be easily determined [21].

Since different κ-light chains or λ-light chains cannot be distinguished on the protein level using antibodies, the existence of colonies producing both isotypes at the same time has been used as a measure for light-chain commitment of the pre-B cells. In colonies derived from pre-B cells from 13- to 15-day-old fetal liver, about 5% of all colonies expressed κ- and λ-light chains simultaneously [21]. Due to the higher probability of successful κ-gene rearrangements over λ-gene rearrangements, many of the 'κ only' colonies will produce different κ-chains at the same time. Using the statistical calculations described in detail by Sauter and Paige [21], about 50% of the clonable pre-B cells are already light-chain-committed at the initiation of culture, whereas the other half give rise to daughter cells which rearrange different light-chain genes in vitro.

The existence of light-chain-mixed colonies raised the question of whether the assay also detects pre-B cells before heavy-chain gene commitment. To approach this question, pre-B cells from a combination of congenic mouse strains (BALB/c × CAL20)F$_1$, have been used, in which the IgM-allotypes can be distinguished by antibodies. B cells from BALB/c mice express IgM of the a-allotype recognized by the monoclonal antibody Bet1 [22], whereas the IgM in CAL20 mice is of the e-allotype recognized by the antibody MB86 [23]. In a 15-day-old fetal liver about 10% of the pre-B cells differentiate into allotype-mixed colonies [24]. Since heavy-chain-uncommitted pre-B cells escape detection when the daughter cells rearrange the same allele, the percentage of uncommitted precursors goes up to 20% in the case of commitment occurring after one cell division. Thus, clonable pre-B cells represent different stages of B cell development, ranging from a totally

uncommitted pre-B cell to a late pre-B cell which is committed to both heavy-
and light-chain genes.

Influence of Growth Factors on the Differentiation of Pre-B Cells

The possibility to distinguish particular developmental stages in normal pre-
B cells formed the basis for experiments to explore the influence of growth factors
on pre-B cells. The above described results have been obtained by using a
heterogeneous layer of adherent cells from fetal liver. Obviously, these cells
provide all the growth signal required for the transition from a heavy-chain-
uncommitted pre-B cell to a surface immunoglobulin-positive B cell.

It should be mentioned that direct cell-to-cell contact between feeder
cells and pre-B cells is prevented in the system, suggesting that soluble growth
factors alone are sufficient to promote pre-B cell differentiation. Nevertheless, it remains to be established whether direct cell-to-cell interactions play a
role in vivo, for example, by providing locally very high concentrations of
growth factors.

In order to examine whether feeder cells could be replaced by already
known growth factors, fetal liver cells were stimulated with a series of
cytokines. Two growth factors, interleukin-3 (IL-3) [25] and colony-stimulat-
ing factor-1 (CSF-1) [26] showed an effect: both factors stimulated the
development of mature B cells from pre-B cells [27]. In order to examine
whether this effect was equivalent to the stimulation by feeder cells, the
developmental stage of the responding pre-B cells was determined. IL-3
stimulated both the generation of heavy-chain-mixed and light-chain-mixed
colonies, indicating that this factor could replace feeder cells. In contrast to
this, upon stimulation by CSF-1, only light-chain-mixed colonies, but not
heavy-chain-mixed colonies, developed [27]. This indicated that under these
conditions only pre-B cells already committed to heavy chain could differentiate into mature B cells. In cultures with both IL-3 and CSF-1, the total
number of stimulated pre-B cells was identical to that in cultures stimulated
with IL-3 alone. This result showed that the pre-B cells responding under the
two sets of growth conditions were a highly overlapping population of cells
belonging to the same differentiation pathway. However, pre-B cells which
had not yet undergone heavy-chain gene commitment needed a growth signal
provided upon stimulation with IL-3, but not with CSF-1. This is the first
indication that pre-B cells at different stages of development also differ in
their growth requirements.

In order to distinguish between a direct and an indirect effect, pre-B cells were purified on the basis of expression of the B lineage surface antigen B220 [28, 29] and then cultured with CSF-1 or IL-3. Under these circumstances, no differentiation into mature B cells could be observed [30, 31]. However, if small numbers of pre-B-cell-depleted fetal liver cells were added to these cultures, the effect of these growth factors on pre-B cells could be restored. This result indicated that the action of IL-3 and CSF-1 was indirect via accessory cells which, upon stimulation by IL-3 or CSF-1, produced pre-B-cell-active growth factors. Efforts to replace IL-3- and CSF-1-stimulated accessory cells by the interleukins 1, 2, 4, 5 or 6, alone or in different combinations, have not been successful. From these findings it can be concluded that the differentiation of pre-B cells requires novel differentiation factors which stimulate pre-B cell differentiation in a stage-specific way.

References

1   Paige CJ: Surface Ig-negative B cell precursors detected by formation of antibody-secreting colonies in agar. Nature 1983;302:711–713.
2   Bradley TR, Metcalf D: The growth of mouse bone marrow cells in vitro. Aust J Exp Biol Med Sci 1966;44:287–299.
3   Pluznik DH, Sachs L: The induction of clones of normal mast cells by a substance from conditioned medium. Exp Cell Res 1966;43:553–563.
4   Stephenson JR, Axelrad AA, McLeod DL, Shreeve MM: Induction of colonies of hemoglobin-synthesizing cells by erythropoietin in vitro. Proc Natl Acad Sci USA 1971;68:1542–1546.
5   Metcalf D, MacDonald HR, Odartchenko N, Sordat B: Growth of mouse mega-karyocyte colonies in vitro. Proc Natl Acad Sci USA 1975;72:1744–1748.
6   Metcalf D, Johnson GR, Mandel TE: Colony formation in agar by multipotential hemopoietic cells. J Cell Physiol 1979;98:401–420.
7   Jacobs SW, Miller RG: Characterization of in vitro T-lymphocyte colonies from spleen of nude mice. J Immunol 1979;122:582–584.
8   Metcalf D: Hemopoietic Colony Stimulating Factor. Elsevier, Amsterdam, 1984.
9   Metcalf D, Nossal GVJ, Warner NL, Miller JFAP, Mandel TE, Layton JE, Gutman GA: Growth of B lymphocyte colonies in vitro. J Exp Med 1975;142:1534–1549.
10 Metcalf D, Warner NL, Nossal GJV, Miller JFAP, Shortman K, Rabellino E: Growth of lymphocyte colonies in vitro from mouse lymphoid organs. Nature 1975;255:630–631.
11 Metcalf D, Wilson JW, Shortman K, Miller JFAP, Stocker J: The nature of the cells generating B-lymphocyte colonies in vitro. J Cell Physiol 1976;88:107–116.
12 Kincade PW, Ralph P, Moore MAS: Growth of clones in semi-solid culture is mitogen dependent. J Exp Med 1976;143:1265–1269.
13 Paige CJ, Skarvall H: Plaque formation by B cell colonies. J Immunol Methods 1982;52:57–61.

14   Metcalf D: Role of mercaptoethanol and endotoxin in stimulating B lymphocyte colony formation in vitro. J Immunol 1976;116:635–638.
15   Burrows P, Lejeune M, Kearney JF: Evidence that murine pre-B cells synthesize μ heavy chains but no light chains. Nature 1979;280:838–841.
16   Maki R, Kearney J, Paige C, Tonegawa S: Immunoglobulin gene rearrangement in immature B cells. Science 1980;209:1366–1369.
17   Tonegawa S: Somatic generation of antibody diversity. Nature 1983;302:576–581.
18   Alt FW, Yancopoulos GD, Blackwell TK, Wood C, Thomas E, Boss M, Coffman R, Rosenberg N, Tonegawa S, Baltimore D: Ordered rearrangement of immunoglobulin heavy chain variable region segments. EMBO J 1984;3:1209–1219.
19   Reth MG, Alt FW: Novel immunoglobulin heavy chains are produced from DJH gene segment rearrangements in lymphoid cell. Nature 1984;312:418–423.
20   Reth MG, Ammirati P, Jackson S, Alt FW: Regulated progression of a cultured pre-B-cell line to the B-cell stage. Nature 1985;313:353–355.
21   Sauter H, Paige CJ: Detection of normal B-cell precursors that give rise to colonies producing both κ and λ light immunoglobulin chains. Proc Natl Acad Sci USA 1987; 84:4989–4994.
22   Sauter H, Paige CJ: Differentiation of murine B-cell progenitors in agar culture: Determination of the developmental potential of clonable pre-B cells. Curr Top Microbiol Immunol 1987;135:65–74.
23   Kung JT, Sharrow O, Sieckmann DG, Lieberman R, Paul WE: A mouse IgM allotypic determinant (IgH 6-5) recognized by a monoclonal rat antibody. J Immunol 1981;127:873–876.
24   Nishikawa SI, Sasaki Y, Kina T, Amagai T, Katsura Y: A monoclonal antibody against Igh 6-4 determinant. Immunogenetics 1986;23:137–139.
25   Sauter H, Paige CJ: B cell progenitors have different growth requirements before and after immunoglobulin heavy chain commitment. J Exp Med 1988;168:1511–1516.
26   Ihle JN, Keller J, Oroszean S, Henderson LE, Copeland TD, Fitch F, Prystowsky MD, Goldwasser E, Schrader JW, Palaszynsky E, Dy M, Lebee B: Biologic properties of homogeneous interleukin 3. J Immunol 1983;131:282–287.
27   Stanley ER, Heard PM: Factors regulating macrophage production and growth. J Biol Chem 1977;252:4305–4310.
28   Scheid MP, Landreth KS, Tung J-S, Kincade PW: Preferential but nonexclusive expression of macromolecular antigens on B-lineage cells. Immunol Rev 1982;69: 141–159.
29   Kincade PW, Lee G, Watanabe T, Sun L, Scheid M: Antigens displayed on murine B-lymphocyte precursors. J Immunol 1981;127:2262–2268.
30   Paige CJ, Gisler RH, McKearn JP, Iscove NN: Differentiation of murine B cell precursors in agar culture: Frequency, surface marker analysis and requirements for growth of clonable pre-B cells. Eur J Immunol 1984;14:979.
31   Paige CJ, Skarvall H, Sauter H: Differentiation of murine B cell precursors in agar culture: II. Response of precursor-enriched populations to growth stimuli and demonstration that the clonable pre-B cell assay is limiting for the B cell precursor. J Immunol 1985;134:3699.

Helmut Sauter, PhD, Unité de Génétique Somatique, Institut Pasteur,
25, rue du Dr. Roux, F–75724 Paris Cedex 15 (France)

Sorg C (ed): Molecular Biology of B Cell Developments.
Cytokines. Basel, Karger, 1990, vol 3, pp 132–153

The Scid Mouse Mutant:
Biology and Nature of the Defect

Walter Schuler

Basel Institute for Immunology, Basel, Switzerland

The various types of blood cells, e.g. cells of the myeloid and lymphoid lineages, derive from a pluripotent hematopoietic stem cell. The source of hematopoietic stem cells and the site of blood cell development is fetal liver or adult bone marrow. Blood cell differentiation is under the control of a number of genes, and a mutation in any one of these genes can disrupt differentiation and arrest the cells at a certain stage, resulting in a defect in blood cell development. The analysis of such developmental defects has contributed much to our knowledge of the mechanism of hematopoietic stem cell differentiation and of the individual steps in the differentiation pathway. One such mutation which was first reported in 1983 [1] and has since experienced an exponentially growing interest, is the mouse mutation *scid*. This mutation adversely affects early lymphoid differentiation. Mice homozygous for *scid* lack functional T and B lymphocytes, resulting in a disease syndrome first described in human infants as 'Swiss-type agammaglobulinemia' [2], a form of severe combined immune deficiency or scid, for short. It is thought that the mouse mutation *scid* leads to an inability to assemble the genes correctly that encode antigen-specific receptors of T and B lymphocytes [3]. The purpose of this article is to briefly review the current status of our knowledge about the scid mouse mutant, with special emphasis on the nature of the defect.

Occurrence of the scid Mutation

The *scid* mutation was discovered in the early 1980s by Bosma et al. [1], at the Fox Chase Cancer Center in Philadelphia. This mutation occurred spontaneously in the immunoglobulin congenic BALB/c inbred mouse strain C.B-17 [BALB/c × C57BL/Ka-*Igh*-1^b/ICR (N17 F34)]. C.B-17 mice differ from BALB/c mice only by a portion of chromosome 12, which comes from the C57BL/Ka strain and includes the C57BL/Ka immunoglobulin heavy-chain genes *(Igh)* as well as other genes closely linked to *Igh*. *scid* segregates as a single autosomal, recessive gene [1]; it maps to the centromeric end of chromosome 16 where it is closely linked to the recessive coat color marker mahoganoid *(md)* [4].

Most of the relevant published work has been done with C.B-17 mice homozygous for the *scid* mutation (C.B-17 Icr-*scid/scid*), which here are simply referred to as scid mice, but the *scid* mutation is also being successfully introduced into other mouse inbred strains (e.g. BALB/c [5]) by selective breeding.

Effects of the scid *Mutation: General Phenotype of Scid Mice*

Scid Mice Lack Functional T and B Lymphocytes
While the bone marrow of scid mice appears not to be morphologically different from that of their normal co-isogenic counterparts, there are other striking histological differences [1, 6]. Scid mice are lymphopenic; their thymus is very small (between 1 and 10% of normal size), consisting of a rudimentary medulla without cortex [1, 6]. Most of the scid thymocytes express Thy-1 and the smaller chain of the interleukin-2 receptor, while CD3, CD4 and CD8 are virtually undetectable on thymocytes from young adult scid mice [1, 7, 8]. Lymph nodes are tiny and usually only detectable histologically through retention of their basic stromal structure and sinus pattern; Peyer's patches and bronchial lymphocytic foci are undetectable [6]. The spleens of scid mice are usually smaller than normal and, like their lymph nodes, are virtually devoid of lymphoid cells [1, 6–8]. Also, scid mice lack dendritic Thy-1$^+$ epidermal cells in their skin [9].

One trait of scid mice, the one that led to their discovery, is their lack of serum immunoglobulin (<1 µg/ml) [1]. Scid mice cannot mount an immune response to T-cell-independent or T-cell-dependent antigens [1] and are unable to reject allogeneic grafts [1, 10]. Scid splenocytes do not proliferate in vitro in response to allogeneic cells and cytotoxic activity cannot be generated in mixed lymphocyte cultures [11]. The T cell and B cell mitogens concanavalin A and lipopolysaccharide fail to stimulate proliferation of splenocytes from scid mice [11].

The above observations indicate that scid mice lack mature, functional T and B lymphocytes. With respect to the precursor cell level, pre-B cells, that is cells expressing the B lineage surface marker B220 and cytoplasmic IgM heavy chains (Cμ), are virtually undetectable in bone marrow from scid mice [1, 8, 12].

The scid *Mutation Results in a Stem Cell Defect*
Scid mice can be cured of their lymphoid deficiency by transplantation of normal hematopoietic stem cells [1, 6, 13]. Transplanted bone marrow or fetal liver cells from normal immunoglobulin congenic donor mice readily differentiate into lymphocytes and repopulate the lymphatic organs of scid mice [1, 6]; within 4–6 weeks after cell transfer, production of immunoglobulin of the donor allotype is found, while immunoglobulin of the host (scid) allotype remains undetectable [1]. Reconstitution of untreated scid mice is variable and often incomplete [6, 13], but full reconstitution can be achieved by sublethally irradiating adult scid recipients [13], or using newborn scid mice [14]. In contrast, lethally irradiated normal recipients injected with bone marrow or fetal liver cells from scid donors do not become repopulated with lymphocytes of the donor (scid) allotype, while myeloid cells of scid origin are easily detectable (see below) [1].

These experiments demonstrate that the *scid* mutation does not impair the development and function of ancillary cells which support lymphoid differentiation (e.g. stroma cells of fetal liver or of bone marrow), but that the resulting defect is rather an intrinsic hematopoietic stem cell defect.

The scid *Mutation Affects a Process Unique to Lymphoid Differentiation*
In contrast to the lymphoid lineage, differentiation and function of cells of the myeloid lineage appear not to be affected by the *scid* mutation. The number of in vitro myeloid colony-forming units is comparable in scid and normal bone marrow [11]. Hematocrit is in the normal range, and erythrocytes of lethally irradiated hosts reconstituted with C.B-17 *scid/scid* bone marrow were found to be entirely of the donor C.B-17 (H-2^d) haplotype [1]. Granulocytic hyperplasia, usually accompanied by marked megakaryocytosis, is frequently found in spleens of scid mice with pyogenic infections [6]. Furthermore, the function of antigen-presenting cells is normal [15], and microbial infection (e.g. with *Listeria monocytogenes*) results in a T-cell-independent activation of scid macrophages associated with enhanced macrophage Ia antigen expression [16–18]. Natural killer (NK) cell activity is

normal in scid mice [19]; scid and normal bone marrow contain equivalent numbers of transplantable NK progenitor cells [20]. NK cells are the likely source of interferon-γ which mediates Ia antigen expression of scid macrophages after *Listeria* infection [17, 18]. Interestingly, irradiated scid mice are able to reject bone marrow allografts, a capacity which has also been attributed to NK cells [21].

In summary, since lymphoid and myeloid cell lineages both derive from a common hematopoietic stem cell [22–25], these observations indicate that the *scid* mutation affects a critical process unique to lymphoid differentiation.

Nature of the Scid Defect: A Defective VDJ Recombinase System?

Rearrangement of Antigen Receptor Genes: A Brief Introduction

A crucial process and hallmark of lymphoid differentiation is the rearrangement of genes that code for antigen-specific receptors expressed on the cell surface of immunocompetent lymphocytes. The B cell receptors, i.e. immunoglobulins, consist of heavy (Igh) and light (Igl) chains. The T cell receptors (TCRs) are composed of either α- and β-chains, or of γ- and δ-chains. Early in lymphocyte development, distinct and separate germ line DNA segments, i.e. V, D and J, or V and J, are assembled in recombinatorial fashion to form the coding sequences for the antigen-binding, variable regions of immunoglobulins and TCRs [for review, see for instance ref. 26, 27]. This procedure is one of the basic ways in which the immune system generates a high degree of diversity of its antigen-recognizing elements. VDJ rearrangement appears to proceed in an ordered fashion such that D to J recombination occurs first, followed by joining of a V gene segment to DJ. In developing B cells, rearrangement of *Igh* genes precedes that of *Igl* genes. Likewise, in developing T cells, the *Tcr-δ*, *Tcr-γ* and *Tcr-β* genes are rearranged prior to *Tcr-α* genes. Rearrangement is thought to be mediated by a site-specific recombination system (the so-called VDJ recombinase system), which recognizes highly conserved heptamer-nonamer DNA sequence motifs (recognition sequences) flanking the V, D and J gene segments of both TCR and immunoglobulin genes. Both T and B lymphocytes make use of the same VDJ recombinase system [28]. This recombinase system has not been characterized yet and the underlying mechanism of V(D)J assembly is only partly understood. A simplified schematic view of V(D)J recombination is given in figure 1a. As an initial step, the two gene segments to be recombined

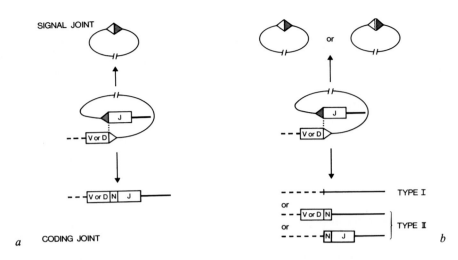

Fig. 1. Schematic representation of the process of V(D)J recombination and the products formed in normal lymphocytes *(a)* and lymphocytes from scid mice *(b)*. The open and dotted triangles represent the heptamer-nonamer recognition sequences flanking V, D and J gene segments. The wavy lines drawn in scid signal joints indicate deletions in the heptamer-heptamer border region.

(e.g. a given D and J, or V and J) are juxtaposed. It is thought that this is then followed by cleavage of the DNA strands at both coding segment/signal sequence junctions. Two products are subsequently formed. The heptamers of the signal sequences are fused back-to-back, forming the signal joints, and the two coding segments are joined, forming the coding joints. This process generally results in a circularization and deletion of the intervening DNA, although rearrangement by inversion has also been observed for certain antigen receptor genes. In this case, the signal joints are retained on the same chromosome. The formation of signal joints is usually very precise while the ends of the coding segments are often modified, i.e. a few basepairs are lost and/or added (N region addition) before they are ligated. Synthesis, assembly and cell surface expression of the antigen receptor will occur only if the respective V(D)J recombination at two different loci (i.e. *Igh* and *Igl*, *Tcr-α* and *Tcr-β*, *Tcr-γ* and *Tcr-δ*) is productive, that is, if each gene segment is in the same translational reading frame after rearrangement and if there are no in-phase nonsense codons. If a productive rearrangement does not occur on either allele of an antigen receptor gene, then the affected developing lymphocyte will be nonfunctional for lack of an antigen receptor and will probably not expand further.

As will be shown in the following, inability to undergo *productive* rearrangement of antigen receptor genes seems to be the explanation for the lack of functional T and B lymphocytes in scid mice. The hypothesis that the *scid* mutation affects a component of the VDJ recombinase system common to both T and B cells [3] will be discussed.

T- and B-Lineage-Committed Cells Do Arise in Scid Mice but Appear not to Rearrange Their Antigen Receptor Genes

Prior to rearrangement, antigen receptor genes become transcriptionally active and are transcribed from germ line configuration. Some of these germ line transcripts appear to reflect early lineage commitment of developing lymphocytes, since germ line transcripts of immunoglobulin κ-light-chain constant-region (*Igl*Cκ) and heavy-chain variable-region (*Igh*V$_H$) genes are exclusively found in B lineage cells [29, 30], while germ line *Tcr*-β transcription is restricted to T lineage cells [31, 32]. Germ line transcripts of antigen receptor genes are also thought to reflect a change in the chromatin structure and an 'opening' of the respective antigen receptor genes which renders them accessible not only to transcription factors, but also to the VDJ recombinase system [30].

Although T- and B-lineage-committed cells were undetectable by either serological or functional assays (see above), we found lineage-specific transcripts of unrearranged antigen receptor genes in hematopoietic tissues of scid mice [5]. In particular, in 16-, 17- and 18-day fetal liver as well as in adult bone marrow, germ line transcripts of the *Igh*Cμ gene, of the *Igh*V$_H$558 gene family and of the *Igl*Cκ gene were observed. Likewise, a high level of germ line *Tcr*-γ transcripts and a low level of germ line *Tcr*-β transcripts were found in thymocytes of young adult scid mice while *Tcr*-α transcripts were undetectable [5]. The latter transcription pattern resembles that seen in 15-day fetal thymocytes from normal mice. In addition, transcripts of the B-lineage-specific λ$_5$ gene were found in scid fetal liver and adult bone marrow while transcripts of the T-lineage-specific T3δ gene were found in scid thymocytes [5]. These results indicate that T- and B-lineage-committed cells do indeed develop in scid mice.

Further evidence for the existence of early T lineage cells is the development of thymic lymphomas in some 15–20% of scid mice. These tumors arise spontaneously; they originate in the thymus, express T-cell-specific markers on their surface and are classified as T cell lymphomas [6]. The high incidence of T cell lymphomas in scid mice is surprising since other types of tumors are rare in these mice (e.g. B cell lymphomas have not been observed [6]) and since their normal C.B-17 (and BALB/c) counterparts do not spontaneously develop T cell lymphomas. It appears that the development of T cell lymphomas is somehow a consequence of the *scid* mutation. There is also additional evidence for early scid B cells in that cells from adult scid bone marrow or from fetal liver can be transformed by Abelson murine leukemia virus (A-MuLV) [3, 33], a virus known to transform early B cells. In fact, A-MuLV infection of adult scid bone marrow yielded comparable numbers of transformants as that of bone marrow from normal mice [33]. This result indicates that scid mice have normal numbers of early B cell precursors.

The observation of germ line transcripts from antigen receptor genes further suggests that, in the committed scid cells, the respective antigen receptor genes are accessible to the VDJ recombinase system; i.e. the developing scid lymphocytes seem to reach the stage at which antigen receptor genes normally undergo rearrangement. However, we could not demonstrate such rearrangement [3, 5]. For example, the antigen receptor gene transcripts found in scid hematopoietic tissues corresponded exclusively to germ line genes, while transcripts of rearranged *Ig* or *Tcr* genes were undetectable [5]. This is consistent with results obtained by Southern blot analyses. Nongerm-line fragments hybridizing to *Tcr*-β or *Tcr*-γ probes could easily be detected in DNA from normal thymocytes, indicating rearrangement of these TCR genes. In contrast, the respective hybridization pattern of DNA from scid thymocytes was indistinguishable from that obtained with germ line (liver) DNA [3]. Similarly, using *Igh*-specific DNA probes, rearrangement of *Igh* genes was found in 4 out of 15 hybridomas generated from bone marrow cells from normal mice while no rearrangement could be detected in any of 40 hybridoma lines from scid mice [3].

Taken together, the results presented in this section indicate that lymphoid development is not totally blocked in scid mice. Differentiation proceeds beyond lineage commitment, and the effects of the *scid* mutation do not become manifest until the developing scid lymphocytes reach the stage at which antigen receptor genes would normally undergo rearrangement.

Antigen Receptor Gene Rearrangement Is Defective in
Transformed Lymphoid Cells from Scid Mice

As we first reported in 1986 [3], a completely different picture of antigen receptor gene rearrangement is seen in DNA from transformed scid lym-

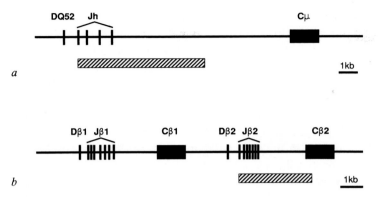

a

b

Fig. 2. Schematic representation of the organization of the *Igh* gene *(a)* and *Tcr*-β chain genes *(b)*. (Further D$_H$ segments are located 10–80 kilobases (kb) upstream from the Jh cluster.) The hatched bars indicate the range of J-associated deletions at the respective loci as revealed by Southern blot analyses of DNA from A-MuLV-transformed scid bone marrow cell lines and from spontaneous scid T cell lymphomas, respectively [from ref. 3, 34, and W.S., unpubl. observ.].

phoid cells, i.e. from spontaneous scid T cell lymphomas and A-MuLV-transformed scid bone marrow cells. Southern blot analysis of DNA from 9 A-MuLV-transformed scid cell lines revealed rearrangement of both *Igh* alleles in all cell lines studied. The striking finding was, however, that 80% of the *Igh* alleles had deleted the entire cluster of J$_H$ coding segments. The deletions frequently extended to a variable extent into the flanking region downstream from the J$_H$ cluster, but the Cμ gene was always retained (schematically indicated in fig. 2a). Further probing with various *Igh*-specific DNA probes demonstrated that, on alleles which had incurred a deletion of the entire J$_H$ cluster, DNA sequences upstream of D$_H$ coding segments were juxtaposed to the Cμ gene. This result indicated that the observed J$_H$-associated deletions resulted from attempted site-specific D$_H$ to J$_H$ recombinations and not from other types of recombination such as chromosomal translocations or viral integration. Loss of J$_H$ coding segments, which was not found to accompany D$_H$ to J$_H$ rearrangements in A-MuLV-transformed cell lines derived from normal mice, certainly renders the affected *Igh* gene nonfunctional. Thus, a high proportion of *Igh* gene rearrangements appeared to be defective in A-MuLV-transformed B lineage cells from scid mice [3]. These results were later confirmed and extended in several laboratories [34–36] by DNA cloning and sequencing of rearranged *Igh* genes from A-MuLV-transformed scid cell lines. These studies revealed, in general, two types of defective recombination which are schematically presented in figure 1b. In the extreme case (type I in fig. 1b), both the D_H and the J_H coding segment involved in the attempted site-specific recombination were deleted. However, there were also rearrangements that appeared normal with respect to one coding segment (type II in fig. 1b), in that one of the involved coding segments (either D_H or J_H) was properly cut and modified (N region addition) at the coding segment/signal sequence junction and ligated to noncoding DNA sequences in the respective upstream or downstream flanking region of the other coding segment. This rearrangement resulted in the deletion of one of the recombination partners. Transitional forms of type I and type II rearrangement, where the deletions extended considerably into the remaining coding segment without eliminating it entirely, have also been observed.

Quite analogous results to those described for the A-MuLV-transformed scid B lineage cells were observed with transformed scid T lineage cells. Southern blot analysis of DNA from 12 spontaneous scid T cell lymphomas revealed that, in all cell lines, rearrangement occurred on both alleles of all Tcr-β and Tcr-γ genes while Tcr-α genes were not rearranged [3; W.S. et al., submitted]. This indicated that these transformed scid cell lines represent immature T lymphocytes. Similar to the situation found for A-MuLV-transformed cell lines from scid mice, about 60% of the Tcr-β genes lacked the entire $J\beta_2$ gene segment cluster [3] including various degrees of downstream DNA (fig. 2b). Again, it could be shown by Southern blotting [3] and confirmed by DNA cloning and sequencing [W.S. et al., in preparation] that the $J\beta_2$-associated deletions resulted from attempted site-specific $D\beta$ to $J\beta$ recombinations. $J\beta_2$-associated deletions accompanying $D\beta$ to $J\beta_2$ recombinations are clearly abnormal since they render the affected Tcr-β genes nonfunctional. Such deletions are not found in T lymphocytes or T cell tumors from normal mice [3, 37]. The DNA sequence analysis revealed the same types of defective recombination as described above for D_H to J_H recombination in A-MuLV-transformed scid cell lines, i.e. the attempted $D\beta$ to $J\beta_2$ recombinations resulted in the deletion of either both (type I) or only one coding segment (type II). Type II recombinations, where $D\beta$ is lost but the $J\beta_2$ segment is retained, cannot be detected by Southern blot analysis under the conditions used. Therefore, the frequency of defective $D\beta$ to $J\beta_2$ recombinations is even higher in the transformed scid T lineage cells than suggested by the Southern blot data. The status of $J\beta_1$ gene segments in DNA from these scid T cell lymphomas was not analyzed. The reason for this was that deletion of $J\beta_1/C\beta_1$ is frequently found in normal T cells as a result of a $D\beta_1$ to $J\beta_2$ or $V\beta$ to $DJ\beta_2$ recombination.

Rearrangements of *Igh* and *Tcr*-β V_H gene segments have not been detected in most of the A-MuLV-transformed scid B lineage cells [3, 34, 35] and T cell lymphomas [W.S., unpubl. observ.]. A likely explanation for this is the lack of an acceptor site for a V gene segment since these sites were generally deleted as a result of a preceding attempt to join D and J (type I in fig. 1b). However, when the 5'-end of a recombined D gene segment is left intact (type II in fig. 1b) V to D rearrangements should still be possible. In fact, one likely example of an attempted V_H gene recombination to an aberrant DJ_H segment in a scid A-MuLV-transformed pre-B cell line has been reported [36]. This attempted recombination apparently resulted in the deletion of the respective V_H gene.

Likewise, rearrangement of *Igl* genes, which lack D gene segments, has not been found in most transformed scid B lineage cells studied [3, 34, 35]. However, two rearrangements of *Igl*-κ genes have recently been reported in 2 subclones of A-MuLV-transformed scid cell lines [38]. One rearrangement lacked *Igl*-κ variable ($V\kappa$) genes, the other lacked both $V\kappa$ and $J\kappa$ gene segments. Furthermore, analysis of the status of *Tcr*-γ genes (which also lack D segments) in the 12 spontaneous scid T cell lymphomas reported above revealed that, of 47 rearranged *Tcr*-γ genes, 36 lacked $V\gamma$ and/or $J\gamma$ gene segments [W.S. et al., submitted]. Sequencing of some of these rearranged *Tcr*-γ genes demonstrated the same types of defective recombination as described above for rearrangement of *Igh* and *Tcr*-β genes. Thus, not only is D to J recombination defective in transformed scid lymphoid cells, but also V to J rearrangement.

The deletion breakpoints were in all cases randomly distributed on the *Igh, Tcr*-β and *Tcr*-γ genes, and scattered over several kilobases of DNA in the flanking region of V or D and J gene segments. A common DNA sequence near the novel recombination sites, which might serve as a pseudo-recognition signal, has not been found.

Hypothesis: The VDJ Recombinase System Is Defective in Scid Mice

In attempting to explain our contradictory results, i.e. no detectable antigen receptor gene rearrangements in *nontransformed* cells versus high frequency of defective rearrangements in *transformed* scid lymphoid cells, we proposed [3] that a component of the VDJ recombinase system, common to both T and B lineage cells, might be altered or missing in scid mice. This would result in the inability of developing scid lymphocytes to generate functional antigen receptor genes, and thus in the inability to express an antigen receptor protein on the cell surface. We further assumed that early lymphocytes which have not succeeded in making a productive rearrangement of a critical antigen receptor gene (i.e. *Igh* or *Tcr*-β), and which are therefore of no use to the immune system, are prematurely and rapidly eliminated from lymphopoietic tissues. This would account for the apparent absence of antigen receptor gene rearrangements in scid lymphoid cells and would explain why the detection of such defective cells depends on their immortalization, e.g. by viral transformation. The assumption of a mechanism which senses and eliminates useless, defective lymphoid cells during lymphocyte development does not appear unreasonable, considering that VDJ recombination is highly error-prone even under normal circumstances. On theoretical grounds, two thirds of V(D)J recombinations are nonproductive, i.e. will not result in the synthesis of an antigen receptor polypeptide, because the J segments will not be in the proper translational reading frame [39, 40]. To benefit from a high diversity of its antigen-specific receptors, the immune system seems to put up with a rather large production of nonfunctional cells that have to be removed.

One of the difficulties of the above hypothesis is that it was based on studies of transformed scid lymphocytes where the effects of the transforming agents are not known. However, subsequent observations make it highly unlikely that the defective rearrangements result from transformation. First, it could be shown that long-term culture conditions allowed the differentiation of scid bone marrow cells into cells expressing the B lineage marker B220 [41, 42]. Although IgM was not (or only rarely) synthesized, the cultured cells had rearranged their *Igh* genes. Cloning and sequencing of rearranged *Igh* genes from in vitro cultured scid bone marrow cells demonstrated that these *Ig* rearrangements were defective; the same types of abnormal deletions as described above for *Igh* gene rearrangement in A-MuLV-transformed scid cells (schematically presented in fig. 1b) were found [43]. The interesting question remains as to why the proposed screening and elimination mechanism is not operational in vitro. Second, most recently it has been reported that, while rearrangements of *Tcr*-γ and *Tcr*-β genes were undetectable by means of Southern blot analysis of DNA from scid thymocytes, rearrangement of *Tcr*-δ genes could be readily detected in the same DNA [44]. Interestingly, DNA from the thymus of individual scid mice all displayed the same uniform pattern of 3 nongerm-line *Tcr*-δ-containing restriction fragments. Given the random extent of the abnormal deletions accompanying V(D)J recombinations in transformed T and B lineage cells from scid mice, this result was unexpected. Although these *Tcr-δ* gene rearrangements have not yet been characterized in detail, this observation provides the first direct evidence of antigen receptor gene rearrangement in nontransformed, developing scid lymphocytes. Why such cells can accumulate and thus become detectable in scid thymus is unclear. One explanation is that developing T lymphocytes containing nonfunctional *Tcr-δ* genes are not under the control of the above postulated elimination mechanism. This seems reasonable since under normal circumstances those cells will have a second chance because they could still productively rearrange *Tcr-α* and *Tcr-β* genes and express an αβ-TCR.

The idea that the *scid* mutation results in a defective VDJ recombinase system [3] is now generally accepted. Recent reports [38, 45, 46] have more precisely defined the effects of the *scid* mutation and have pinpointed the *scid* defect to the step in VDJ recombination at which formation of the coding joint occurs. Studies of VDJ recombination substrates transfected into A-MuLV-transformed pre-B cell lines from scid mice, either as extrachromosomal plasmids [45] or as integrated retroviral vectors [46] demonstrated rearrangement of these substrates which led with normal frequency to the formation of signal joints. In contrast, coding joints could not be detected. Similar observations were made in subclones of 2 A-MuLV-transformed scid pre-B cell lines which had rearranged their endogenous *Igl-κ* genes [38]. This indicates that the scid VDJ recombinase system is still capable of recognizing and binding to the heptamer-nonamer signal sequences, of cleaving the DNA strands at the proper position and of generating signal joints (fig. 1b). The scid enzyme appears to fail, however, in joining the open coding ends. It has been proposed [36] that this might result in double-strand breaks within the chromosome and that the observed aberrant joints (type I and II in fig. 1b) in immature scid lymphocytes might result from secondary processes, e.g. illegitimate recombinations, rescuing the affected cells from the otherwise lethal chromosomal breaks.

The formation of signal joints is always very precise in the recombination process of endogenous *Igl-κ* genes or stably integrated exogenous recombination substrates in A-MuLV-transformed pre-B cells from both scid and normal mice. That is, no loss or addition of nucleotides at the heptamer-heptamer border has been observed [38, 46]. This has been interpreted as a demonstration that the formation of signal joints is mechanistically distinct from coding joint formation [38, 46]. However, this may not be the case. Studies with an extrachromosomal plasmid recombination substrate, which allows screening of a much larger number of signal joints, demonstrated that loss of nucleotides occurred in about 50% of the signal joints formed in scid cells [45]. No such loss could be found in signal joints generated in normal cells [45]. It is not yet clear as to what extent topological differences might be responsible for the contrasting results obtained with chromosomal versus extrachromosomal recombination substrates.

Transgenic Scid Mice Carrying Productively Rearranged Antigen Receptor Genes

As discussed above, there is now considerable evidence that the lack of functional T and B lymphocytes in scid mice is solely due to a defective VDJ recombination process. If this was the case, one would predict that the introduction of functionally rearranged antigen receptor genes into the genome of scid mice should overcome the developmental arrest and allow further differentiation and maturation of scid lymphocytes. Experiments with scid mice transgenic for productively rearranged *Tcr* genes [47] or *Ig* genes [48] have in fact confirmed this prediction.

For technical reasons, transgenes are introduced into the scid mouse genome by selective breeding with established transgenic mouse lines. Introduction of the *Tcr-α* and *Tcr-β* genes encoding a TCR with specificity for the male (HY) antigen in the context of the mouse MHC class I H-$2D^b$ antigens resulted in a dramatic change [47]. The number of thymocytes in female αβ-transgenic scid mice was back to normal levels, with the majority expressing both CD4 and CD8 as well as the transgenic TCR on their surface. Furthermore, in female αβ-transgenic scid mice carrying the proper MHC restriction element (i.e. H-$2D^b$), mature CD4-negative/CD8-positive thymocytes were found. In contrast, the secondary peripheral lymphatic organs were still underdeveloped; lymph nodes contained only about 10% of T lymphocytes when compared to normal αβ-transgenic mice. Immunoglobulin-positive B lymphocytes were not found. The basis for the disparity between the thymus and the periphery is not clear. Repopulation of secondary lymphatic organs might require interaction of various subsets of lymphocytes displaying a diverse spectrum of antigen receptors [47].

Similar results were obtained with scid mice transgenic for the IgM heavy-chain (µ) gene of an NP 4-hydroxy-3-nitrophenyl-specific antibody [48]. In contrast to regular scid mice, bone marrow of µ-transgenic scid mice contained normal numbers of B220-positive cells. Mature surface-IgM-positive cells could not be found, indicating that productive *Igl* gene rearrangement does not occur in these µ-transgenic scid mice.

Development of Functional Lymphocytes in Scid Mice:
Is the scid *Mutation Leaky?*

The block in lymphocyte differentiation is not absolute in scid mice. Depending on their housing conditions (see below), between 2 and 23% of young adult scid mice contain serum immunoglobulin, reaching a frequency of 30–47% among mice 1-year-old and older [49]. Serum-immunoglobulin-positive (Ig$^+$) scid mice not only have functional immunoglobulin-secreting B lymphocytes but also functional T lymphocytes; they have been shown to reject skin allografts [49], and alloreactive T cell clones could be established from those mice [50]. Ig$^+$ scid mice are by no means normal, since in most respects they resemble regular scid mice [49].

The event that leads to the generation of Ig$^+$ scid mice is a somatic event since mice cannot be bred for this condition. It is also a rare event since functional T and B lymphocytes show only restricted clonality. Southern blot analyses suggest that the T cell repertoire is limited to probably not more than 3 clones per mouse [51]. Isoelectric focusing of κ-light chains from sera of 94 Ig$^+$ scid mice revealed that in about half of the mice the serum immunoglobulin corresponded to 4–7 different B cell clones [52]. About 25% had 3 or fewer B cell clones; mice having more than 8–12 different clones were rarely found [52]. Hybridomas generated from spleen cells of a single Ig$^+$ scid mouse used the same V_H, D_H and J_H gene segments, suggesting that the hybridomas were clonally related [53]. This further demonstrates the pauciclonality of functional B lymphocytes in Ig$^+$ scid mice. Individual mice gave rise to unrelated immunoglobulin-producing hybridomas, and no preferential usage of certain V_H, D_H and J_H gene segments has been found [53]. All Igh-chain isotypes can be found in sera of Ig$^+$ scid mice, although individual mice rarely express all of them [49].

The development of Ig$^+$ scid mice appears to be antigen-driven. Most of the monoclonal antibodies derived from Ig$^+$ scid mice react with self-antigens [53]. Some have specificity for certain enterobacteria [53], and the frequency of Ig$^+$ scid mice increases when scid mice are transferred from isolator conditions to a less stringently controlled environment [49]. Recent flow-microfluorimetric analyses [51] have now shown that scid mice harboring functional B lymphocytes (i.e. IgM$^+$/B220$^+$ cells) and functional T lymphocytes (i.e. CD3$^+$ cells) are even more frequent than previously thought. All old scid mice tested (older than 1 year) contained functional T or B lymphocytes, regardless of whether they contained detectable serum immunoglobulin. Interestingly, the lymphocytes found in the periphery resided predominantly in the peritoneal cavity. Both CD5 (Ly-1)-positive and -negative B lymphocytes were found [51].

It is probably safe to say that functional T and B lymphocytes develop at a very low frequency in every scid mouse. These rare lymphocytes most likely die for lack of a further stimulus within a few days unless they encounter the proper antigen and become clonally expanded and thus detectable. For B lymphocytes, clonal expansion and terminal differentiation into immuno-globulin-producing plasma cells will in most cases depend upon the presence of appropriate helper T lymphocytes. This might explain why all scid mice shown so far to contain serum immunoglobulin, or otherwise detectable B lymphocytes, also contained CD3-positive T lymphocytes [51]. In the converse situation, some scid mice were found which contained only T lympho-cytes but no detectable B lymphocytes [51]. However, the failure to detect B lymphocytes could be due to technical limitations and, therefore, an alterna-tive explanation might be that there is an interdependence in the develop-ment of T and B lymphocytes.

Among others, two explanations can be considered to account for the development of functional lymphocytes in scid mice. Both explanations are based on the hypothesis that the result of the *scid* mutation is a defective VDJ recombinase system. The first explanation assumes that the defective recom-binase activity may be on rare occasions somehow 'normalized' in very early scid lymphocytes, allowing normal, conventional rearrangements. The sec-ond explanation postulates that the *scid* mutation is leaky; although most V(D)J rearrangements are defective, residual activity of the scid VDJ recombinase system will still allow correct V(D)J recombinations, albeit at a very low frequency. The latter hypothesis predicts that, in a given scid lymphocyte, one should find both abnormal, *scid*-type deletions as well as normal, conventional recombinations. As discussed below, results have been obtained supporting either one of the two hypotheses.

Consistent with the second hypothesis, 5 out of the 12 spontaneous scid T cell lymphomas already mentioned contain both normal and defective rearrangements. In all 5 cell lines all *Tcr*-β genes and most *Tcr*-γ genes were defectively rearranged, but at least one, in some cell lines two, *Tcr*-γ genes had undergone normal rearrangement [W.S. et al., in preparation]. Likewise, in an alloreactive T cell clone derived from an Ig+ scid mouse, abnormal *scid*-type deletions were found on one *Tcr*-β allele as well as on one of the *Tcr*-γ genes [50]. Since this T cell clone was CD3-positive and antigen-responsive, normal productive recombinations must have also occurred in this cell line giving rise to a functional TCR. In contrast to these results, in 3 other scid-derived CD3-positive alloreactive T cell clones, all *Tcr*-β and *Tcr*-γ gene rearrangements were normal [50]. This finding is consistent with the idea of 'normalization' of the VDJ recombinase system where one would expect to find all V(D)J recombinations in a given lymphocyte to be normal.

The two explanations are not mutually exclusive. The *scid* mutation could be a point mutation or an insertion of a transposable element in either a structural or regulator gene, or in a transcriptional control region. As a result, the affected component of the VDJ recombinase system might either be altered without entirely abolishing its activity, or might be expressed at considerably lower levels than normal, leaving residual activity mediating normal V(D)J recombinations with low efficiency. In contrast to a mutation caused by a deletion, the effects of point mutations or insertions can be reversed ('normalization') by a somatic back mutation, e.g. by a second mutation at the same place or by a mutation occurring elsewhere in the affected gene, compensating the effects of the first mutation. In eukaryotes, the frequency of reversion is estimated to range between 10^{-6} and 10^{-7} per locus per generation. Considering the high turnover of developing lymphocytes, it does not seem unlikely that, despite this low reversion frequency, somatic reversions of the *scid* mutation could occur with increasing probability during the lifetime of a scid mouse.

Does the scid *Mutation Affect a DNA Repair System?*

A controversial issue on which data are currently rather limited is the radiation sensitivity of cells from scid mice. It has been reported that scid myeloid cells and scid fibroblasts are twice as sensitive to X-irradiation as respective normal cells [54]. Based on this result, it has been postulated that the *scid* mutation affects an enzyme system that is involved both in some aspect of VDJ recombination and in repair of some types of radiation-induced DNA damage. However, other authors [46] could not demonstrate an impairment of the capability to repair radiation-induced DNA double-strand breaks in A-MuLV-transformed scid bone marrow cells. Certainly this issue needs further attention.

Outlook

In essence, there is a growing body of evidence that the product of the *scid* locus is involved in V(D)J recombination. Therefore, one reason for the great interest in the *scid* mutation is the expectation that characterization of the *scid* gene might eventually lead to the identification of a protein involved in V(D)J recombination.

However, of much greater interest to most people are the possible applications of scid mice, not only in immunology but also in biology in general as well as in biomedical and pharmaceutical research. It is beyond the scope of this article to review in detail the potential uses of scid mice. Only a few applications are pointed out here.

Due to their lack of functional T and B lymphocytes, scid mice are an exciting model for investigating lymphocyte development. For instance, TCR-transgenic scid mice [47] or allogeneic scid chimeras [55] have been used to study T cell development and selection processes in the thymus. Scid mice are ideally suited as graft recipients in which the differentiation capacity of stem cells from fresh or in vitro cultured bone marrow or fetal liver [56, 57] or of cloned progenitor cells [58] can be assessed. Great expectations accompany attempts to establish xenogeneic scid chimeras, for example scid mice reconstituted with elements of the immune system of humans [59–61] or of agriculturally important animals like cattle or sheep. The hope is to establish a small animal model which would allow studies not possible in the donor, for instance analyzing the spread of viruses (e.g. HIV), testing drugs and vaccines, assessing immunotoxicity, and the like. It appears, however, that simply eliminating the problem of graft rejection is not sufficient; although the transplanted xenogeneic cells seem to survive in scid mice, it has not yet been demonstrated that these cells truly reconstitute the mice, e.g. that the transferred cells retain their function as antigen-responsive cells.

Scid mice appear to be a better experimental animal model than athymic nude mice for the propagation of human tumors, for the evaluation of protocols for immunospecific targeting of cytotoxic agents, and for the evaluation of the effect of lymphocytes from tumor patients on tumor progression [61]. Scid mice can be used to study the possible role of humoral and cellular immune responses in the development of polyarthritis and carditis in the wake of certain bacterial infections like the tick-borne Lyme disease (borreliosis) [62]. This approach might also be applicable to certain other diseases, like diabetes, where autoaggressive lymphocytes are thought to play a role. Finally, scid mice allow the growth of xenogeneic hybridomas of rat and human origin. High yields of monoclonal antibodies, free of contaminating immunoglobulins, can be achieved and the antibodies can easily be purified from serum or ascites fluid of hybridoma-bearing scid mice [63].

Under proper pathogen-controlled conditions, scid mice do remarkably well and have the same life span as normal mice. Homozygous scid mice are fertile and breed very well, their litter sizes being comparable to their normal counterparts. Successful breeding colonies of scid mice are now found at many places in the United States, Europe and Japan, and scid mice will soon be commercially available. This will certainly encourage more investigators to employ the unique features of this fascinating and novel mouse mutant.

Acknowledgments

I am greatly indebted to Dr. Melvin J. Bosma of the Institute for Cancer Research, Fox Chase Cancer Center, Pa., for giving me the opportunity to work in his laboratory on the scid defect. I really enjoyed my time at Fox Chase. I thank Drs. R. DiPauli, J.F. Kaufman and M.V. Wiles for critical reading of my manuscript, and Nicole Schoepflin for her enthusiastic and excellent job of typing it.

The Basel Institute for Immunology was founded and is supported by F. Hoffman-La Roche Ltd., Basel, Switzerland.

References

1   Bosma GC, Custer RP, Bosma MJ: A severe combined immunodeficiency mutation in the mouse. Nature 1983;301:527–530.
2   Hitzig WH, Biro Z, Bosch H, et al: Agammaglobulinämie und Alymphocytose mit Schwund des lymphatischen Gewebes. Helv Paediatr Acta 1958;13:551–585.
3   Schuler W, Weiler IJ, Schuler A, et al: Rearrangement of antigen receptor genes is defective in mice with severe combined immune deficiency. Cell 1986;46:963–972.
4   Bosma GC, Davisson MT, Ruetsch NR, et al: The mouse mutation severe combined immune deficiency (scid) is on chromosome 16. Immunogenetics 1989;29:54–57.
5   Schuler W, Schuler A, Lennon GG, et al: Transcription of unrearranged antigen receptor genes in scid mice. EMBO J 1988;7:2019–2024.
6   Custer RP, Bosma GC, Bosma MJ: Severe combined immunodeficiency (SCID) in the mouse. Am J Pathol 1985;120:464–477.
7   Habu S, Kimura M, Katsuki M, et al: Correlation of T cell receptor gene rearrangements to T cell surface antigen expression and to serum immunoglobulin level in scid mice. Eur J Immunol 1987;17:1467–1471.
8   Hardy RR, Kemp JD, Hayakawa K: Analysis of lymphoid population in SCID mice: Detection of a potential B lymphocyte progenitor population present at normal levels in SCID mice by three color flow cytometry with B220 and S7; in Bosma M, Phillips R, Schuler W (eds): The Scid Mouse: Characterization and Potential Uses. Curr Top Microbiol Immunol. Berlin, Springer, 1989, vol 152, pp 19–25.
9   Nixon-Fulton JL, Witte PL, Tigelaar RE, et al: Lack of dendritic Thy-1+ epidermal cells in mice with severe combined immunodeficiency disease. J Immunol 1987;138: 2902–2905.
10 Bosma GC, Fried M, Custer RP, et al: Evidence of functional lymphocytes in some (leaky) scid mice. J Exp Med 1988;167:1016–1033.

11 Dorshkind K, Keller GM, Phillips RA, et al: Functional status of cells from lymphoid and myeloid tissues in mice with severe combined immune deficiency disease. J Immunol 1984;132:1804–1808.

12 Witte PL, Burrows PD, Kincade PW, et al: Characterization of B lymphocyte lineage progenitor cells from mice with severe combined immune deficiency disease (SCID) made possible by long term culture. J Immunol 1987;138:2698–2705.

13 Fulop GM, Phillips RA: Full reconstitution of the immune deficiency in scid mice with normal stem cells requires low-dose irradiation of the recipients. J Immunol 1986;136:4438–4443.

14 Bosma GC, Gibson DM, Custer RP, et al: Reconstitution of scid mice by injection of varying numbers of normal fetal liver cells into scid neonates; in Bosma M, Phillips R, Schuler W (eds): The Scid Mouse: Characterization and Potential Uses. Curr Top Microbiol Immunol. Berlin, Springer, 1989, vol 152, pp 151–159.

15 Czitrom AA, Edwards S, Phillips RA, et al: The function of antigen-presenting cells in mice with severe combined immunodeficiency. J Immunol 1985;134:2276–2280.

16 Bancroft GJ, Bosma MJ, Bosma GC, et al: Regulation of macrophage Ia expression in mice with severe combined immunodeficiency: Induction of Ia expression by a T cell-independent mechanism. J Immunol 1986;137:4–9.

17 Bancroft GJ, Schreiber RD, Bosma GC, et al: A T-cell independent mechanism of macrophage activation by interferon-γ. J Immunol 1987;139:1104–1107.

18 Bancroft GJ, Schreiber RD, Unanue ER: T cell-independent macrophage activation in scid mice; in Bosma M, Phillips R, Schuler W (eds): The Scid Mouse: Characterization and Potential Uses. Curr Top Microbiol Immunol. Berlin, Springer, 1989, vol 152, pp 235–242.

19 Dorshkind K, Pollack SB, Bosma MJ, et al: Natural killer (NK) cells are present in mice with severe combined immunodeficiency (scid). J Immunol 1985;134:3798–3801.

20 Hackett J Jr, Bosma GC, Bosma MJ, et al: Transplantable progenitors of natural killer cells are distinct from those of T and B lymphocytes. Proc Natl Acad Sci USA 1986;83:3427–3431.

21 Murphy WJ, Kumar V, Bennett M: Rejection of bone marrow allografts by mice with severe combined immune deficiency (scid): Evidence that natural killer cells can mediate the specificity of marrow graft rejection. J Exp Med 1987;165:1212–1217.

22 Wu AM, Till JE, Siminovitch L, et al: Cytological evidence for a relationship between normal hematopoietic colony-forming cells and cells of the lymphoid system. J Exp Med 1968;127:455–463.

23 Abramson S, Miller RG, Phillips RA: The identification in adult bone marrow of pluripotent and restricted stem cells of the myeloid and lymphoid systems. J Exp Med 1977;145:1567–1579.

24 Fialkow PJ, Denman AM, Jacobson RJ, et al: Chronic myelocytic leukemia origin of some lymphocytes from leukemic stem cells. Clin Invest 1978;62:815–823.

25 Mintz B, Anthony K, Litwin S: Monoclonal derivation of mouse myeloid and lymphoid lineages from totipotent hematopoietic stem cells experimentally engrafted in fetal hosts. Proc Natl Acad Sci USA 1984;81:7835–7839.

26 Blackwell TK, Alt FW: Immunoglobulin genes; in Hames BD, Glover DM (eds): Molecular Immunology. Oxford, IRL Press, 1989, pp 1–60.

27 Davis MM: T cell antigen receptor genes; in Hames BD, Glover DM (eds): Molecular Immunology. Oxford, IRL Press, 1989, pp 61–79.

28   Yancopoulos G, Blackwell TK, Suh H, et al: Introduced T cell receptor variable region gene segments in pre-B cells: Evidence that B and T cells use a common recombinase. Cell 1989;44:251–259.

29   Van Ness BG, Weigert M, Coleclough C, et al: Transcription of the unrearranged mouse Cκ locus: Sequence of the initiation region and comparison of activity with a rearranged Vκ-Cκ gene. Cell 1981;27:593–602.

30   Yancopoulos GD, Alt FW: Developmentally controlled and tissue-specific expression of unrearranged V_H gene segments. Cell 1985;40:271–281.

31   Yanagi Y, Yoshikai Y, Leggett K, et al: A human T cell-specific cDNA clone encodes a protein having extensive homology to immunoglobulin chains. Nature 1984;308:145–149.

32   Cook WD, Balaton AM: T-cell receptor and immunoglobulin genes are rearranged together in Abelson virus-transformed pre-B and pre-T cells. Mol Cell Biol 1987;7:266–272.

33   Fulop GM, Bosma GC, Bosma MJ, et al: Early B-cell precursors in scid mice: Normal number of cells transformable with Abelson murine leukemia virus (A-MuLV). Cell Immunol 1988;113:192–201.

34   Kim MG, Schuler W, Bosma MJ, et al: Abnormal recombination of *Igh* D and J gene segments in transformed pre-B cells of scid mice. J Immunol 1988;141:1341–1347.

35   Hendrickson EA, Schatz DG, Weaver DT: The *scid* gene encodes a *trans*-acting factor that mediates the rejoining event of Ig gene rearrangement. Genes Dev 1988;2:817–829.

36   Malynn BA, Blackwell TK, Fulop GM, et al: The *scid* defect affects the final step of the immunoglobulin VDJ recombinase mechanism. Cell 1988;54:453–460.

37   Kronenberg M, Goverman J, Haars R, et al: Rearrangement and transcription of β-chain genes of the T cell antigen receptor in different types of murine lymphocytes. Nature 1985;313:647–653.

38   Blackwell TK, Malynn BA, Pollock RA, et al: Isolation of scid pre-B cells that rearrange kappa light chain genes: Formation of normal signal and abnormal coding joins. EMBO J 1989;8:735–742.

39   Alt FW, Blackwell TK, DePinho RA, et al: Regulation of genome rearrangement events during lymphocyte differentiation. Immunol Rev 1986;89:5–30.

40   Kronenberg M, Siu G, Hood LE, et al: The molecular genetics of the T-cell antigen receptor and the T-cell antigen recognition. Annu Rev Immunol 1986;4:529.

41   Witte PM, Burrows PD, Kincade PW, et al: Characterization of B lymphocyte lineage progenitor cells from mice with severe combined immune deficiency disease (scid) made possible by long term culture. J Immunol 1987;138:2698–2705.

42   Hirayoshi K, Nishikawa S-I, Kina T, et al: Immunoglobulin heavy chain gene diversification in the long-term bone marrow culture of normal mice and mice with severe combined immunodeficiency. Eur J Immunol 1987;17:1051–1057.

43   Okazaki K, Nishikawa S-I, Sakano H: Aberrant immunoglobulin gene rearrangement in scid mouse bone marrow cells. J Immunol 1988;141:1348–1352.

44   Carroll AM, Bosma MJ: Rearrangement of T cell receptor delta genes in thymus of scid mice; in Bosma MJ, Phillips RA, Schuler W (eds): The Scid Mouse: Characterization and Potential Uses. Curr Top Microbiol Immunol. Berlin, Springer, 1989, vol 152, pp 63–67.

45   Lieber MR, Hesse JE, Lewis S, et al: The defect in murine severe combined immune deficiency: Joining of signal sequences but not coding elements in V(D)J recombination. Cell 1988;55:7–16.

46 Weaver D, Hendrickson E: The *scid* mutation disrupts gene rearrangement at the rejoining of coding strands; in Bosma MJ, Phillips RA, Schuler W (eds): The Scid Mouse: Characterization and Potential Uses. Curr Top Microbiol Immunol. Berlin, Springer, 1989, vol 152, pp 77–84.

47 Scott B, Blüthmann H, Teh HS, et al: The generation of mature T cells requires interaction of the αβ T-cell receptor with major histocompatibility antigens. Nature 1989;338:591–593.

48 Fried M, Hardy RR, Bosma MJ: Transgenic scid mice with a functionally rearranged immunoglobulin heavy chain gene; in Bosma MJ, Phillips RA, Schuler W (eds): The Scid Mouse: Characterization and Potential Uses. Curr Top Microbiol Immunol. Berlin, Springer, 1989, vol 152, pp 107–114.

49 Bosma GC, Fried M, Custer RP, et al: Evidence of functional lymphocytes in some (leaky) scid mice. J Exp Med 1988;167:1016–1033.

50 Carroll AM, Bosma GC: Detection and characterization of functional T cells in mice with severe combined immune deficiency. Eur J Immunol 1988;18:1965–1971.

51 Carroll AM, Hardy RR, Bosma MJ: Occurrence of mature B (IgM$^+$/B220$^+$) and T (CD3$^+$) lymphocytes in scid mice. J Immunol 1989;143:1087–1093.

52 Gibson DM, Bosma GC, Bosma MJ: Limited clonal diversity of serum immunoglobulin in leaky scid mice; in Bosma MJ, Phillips RA, Schuler W (eds): The Scid Mouse: Characterization and Potential Uses. Curr Top Microbiol Immunol. Berlin, Springer 1989, vol 152, pp 125–136.

53 Kearney JF, Solvason N, Stohrer R, et al: Pauciclonal B cell involvement in production of immunoglobulin in scid Ig$^+$ mice; in Bosma MJ, Phillips RA, Schuler W (eds): The Scid Mouse: Characterization and Potential Uses. Curr Top Microbiol Immunol. Berlin, Springer, 1989, vol 152, pp 137–147.

54 Phillips RA, Fulop GM: Pleiotropic effects of the *scid* mutation: Effects on lymphoid differentiation and on repair of radiation damage; in Bosma MJ, Phillips RA, Schuler W (eds): The Scid Mouse: Characterization and Potential Uses. Curr Top Microbiol Immunol. Berlin, Springer, 1989, vol 152, pp 11–17.

55 Zinkernagel RM, Rüedi E, Althage A, et al: Thymic selection of H-2-incompatible bone marrow cells in scid mice: Differences in T help for induction of B cell IgG responses versus cytotoxic T cells. J Exp Med 1988;168:1187–1192.

56 Dorshkind K, Denis KA, Witte ON: Lymphoid bone marrow cultures can reconstitute heterogeneous B and T cell-dependent responses in severe combined immunodeficient mice. J Immunol 1986;137:3457–3463.

57 Denis KA, Dorshkind K, Witte ON: Regulated progression of B lymphocyte differentiation from cultured fetal liver. J Exp Med 1987;166:391–403.

58 Palacios R, Kiefer M, Brockhaus M, et al: Molecular, cellular and functional properties of bone marrow T lymphocyte progenitor clones. J Exp Med 1987;166:12–32.

59 McCune JM, Namikawa R, Kaneshima H, et al: The scid-hu mouse: Murine model for the analysis of human hemato-lymphoid differentiation and function. Science 1988;241:1632–1639.

60 Mosier DE, Gulizia RJ, Baird SM, et al: Transfer of a functional human immune system to mice with severe combined immunodeficiency. Nature 1988;335:256–259.

61 Bankert RB, Umemoto T, Sugiyama Y, et al: Human lung tumors, patients' peripheral blood lymphocytes and tumor infiltrating lymphocytes propagated in scid mice; in Bosma MJ, Phillips RA, Schuler W (eds): The Scid Mouse: Characterization

and Potential Uses. Curr Top Microbiol Immunol. Berlin, Springer, 1989, vol 152, pp 201–210.

62   Schaible UE, Kramer MD, Museteanu C, et al: The severe combined immunodeficiency (scid) mouse: A laboratory model for the analysis of Lyme arthritis and carditis. J Exp Med 1989;170:1427–1432.

63   Ware CF, Domato NJ, Dorshkind K: Human, rat or mouse hybridomas secrete high levels of monoclonal antibodies following transplantation into mice with severe combined immunodeficiency disease (SCID). J Immunol Methods 1985;85:353–361.

Note Added in Proof

The following reference to work cited as submitted is now published: Schuler W, Schuler A, Bosma MJ: Defective V-to-J recombination of T cell receptor γ chain genes in scid mice. Eur J Immunol 1990;20:545–550.

Walter Schuler, Basel Institute for Immunology,
Grenzacherstrasse 487, CH–4005 Basel (Switzerland)

Sorg C (ed): Molecular Biology of B Cell Developments.
Cytokines. Basel, Karger, 1990, vol 3, pp 154–169

The I.29 Mouse B Cell Lymphoma:
Regulation of the Antibody Class Switch

Janet Stavnezer, Penny Shockett, Yi-Chaung Lin,
Kamal U. Saikh, Minzhen Xu

Department of Molecular Genetics and Microbiology, University of
Massachusetts Medical School, Worcester, Mass., USA

We review here our studies of the mouse B lymphoma, I.29, which have
provided information on many aspects of immunoglobulin class switching
including the mechanism of switch recombination, the regulation of choice
of isotype by lymphokines and the mechanism of induction of the class
switch. The I.29 lymphoma arose spontaneously as an ascites tumor in the
peritoneal cavity of an I/St mouse [1]. The first indication that this lymphoma could be a model for heavy-chain switching was the finding that it
consisted of cells expressing surface-bound IgM(λ) or IgA(λ) with the identical idiotype [2]. By DNA blotting experiments, karyotype analyses and
nucleotide sequencing of cloned heavy-chain genes, we then learned that the
IgM$^+$ cells contain two heavy-chain chromosomes, bearing two rearranged
μ-genes: an expressed μ-gene which has undergone VDJ_4 recombination and
a nonexpressed μ-gene which has undergone DJ_2 recombination [3, 4]. The
heavy-chain constant-region (C_H) genes 3' to the Cμ genes are present in a
germ line configuration on both chromosomes in the IgM$^+$ cells. (See figure 1
for a diagram of the structure of the expressed VDJ Cμ and C_H gene loci in
IgM$^+$ cells.) In the IgA$^+$ cells of the tumor, both μ-genes have been deleted,
having undergone switch recombination events between μ-switch (Sμ) and
Sα sequences on the expressed heavy-chain chromosome, and between Sμ
and Sγ_3 sequences on the nonexpressed chromosome. These DNA recombinations resulted in deletion of all the C_H genes except Cα from the expressed
chromosome, and deletion of the μ- and δ-genes from the nonexpressed
chromosome. The C_H genes 3' to the Cγ_3 gene are in germ line configuration
on the nonexpressed chromosome. The V_H gene segments expressed in the

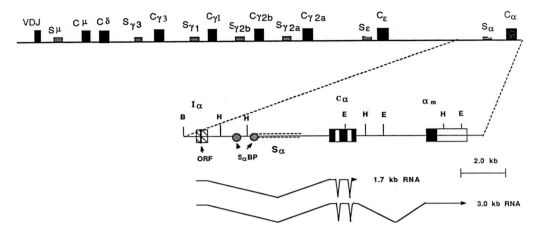

Fig. 1. Diagram (not to scale) of mouse immunoglobulin heavy-chain expressed VDJ Cμ and C$_H$ gene locus in IgM$^+$ cells and expanded map (to scale) of unrearranged Cα gene locus with transcription maps for germ line α-RNAs indicated below. The location of the Iα exon and the 43-amino-acid ORF within this exon are indicated. The location of two binding sites for Sα-BP is indicated. Sα indicates the location of the tandem repeats of the Sα region. Restriction sites noted are: B = BamHI; H = HindIII; E = EcoRI.

IgM$^+$ and IgA$^+$ cells have identical nucleotide sequences and are completely unmutated relative to the germ line sequence. Thus, induction of heavy-chain switching does not activate the somatic mutation process in these cells [4].

The structure of the heavy-chain genes in the IgM$^+$ and IgA$^+$ cells indicated that the IgA$^+$ cells had been derived from the IgM$^+$ cells by heavy-chain switching, but not whether this event was ongoing or had already occurred at a previous time. To determine if the IgM$^+$ cells could undergo heavy-chain switching, IgM$^+$ cells were purified on the fluorescence-activated cell sorter, adapted to culture and cloned [5]. These lines generally remain ≥99% IgM$^+$ when maintained in culture. To attempt to induce heavy-chain switching, I.29μ cells were treated with lipopolysaccharide (LPS) for 4 days. Cells which contained IgA in their cytoplasm were detected about 3 days after addition of LPS by immunofluorescence microscopy of fixed cells. In all experiments discussed below, we assay switching by immunofluorescence microscopy of ethanol-fixed cells. Most of the IgA$^+$ cells at this early time after addition of LPS also contain IgM in their cytoplasm [5]. The double-staining cells must be derived by isotype switching from the IgM$^+$ cells, and not by outgrowth of any IgA$^+$ cells that could be contaminating the IgM$^+$ cell lines, since we had demonstrated that IgA$^+$ cells present in the I.29 tumor have deleted Cμ genes from both of their heavy-chain chromosomes, and thus cannot synthesize IgM [3].

We have performed a number of experiments to search for other molecules, besides LPS, which would induce switching [5]. The greatest amount of switching was consistently induced by LPS (10–50 μg/ml) and a monoclonal anti-I.29 idiotype (Id) antibody [6] when added to the cells for 4 days. For example, 8 days after I.29μ cells were treated with LPS plus purified anti-Id (5 μg/ml), the cultures contained 14% IgA$^+$ cells. The culture treated with LPS alone contained 8% IgA$^+$ cells. Purified anti-Id antibody in the absence of LPS did not induce switching. Thus, anti-Id antibody and LPS act synergistically to induce switching.

Treatment of I.29μ cells with various anti-IgM antibodies did not generally augment the LPS-induced switching. However, these antibodies decreased the viability of the cells. LPS is always required to induce switching. I.29μ cells switch predominantly to IgA and much less frequently (10- to 10^2-fold) to IgE or to IgG2a. In one experiment each, a switch to IgG1 or to IgG3 occurred. We have shown by restriction enzyme mapping and by nucleotide sequencing that switching to IgA and to IgE is accompanied by DNA recombination within the tandem repeats of the switch regions, resulting in deletion of the DNA lying between these sites [5]. The switch to IgG2a is also effected by a DNA recombination, although the site of recombination has not been mapped [J.S., unpubl. data]. We have not studied the structure of the expressed γ$_1$- or γ$_3$-genes. In cells which have switched in culture, the nonexpressed chromosome often has deleted portions of Sμ sequences but retained the Cμ gene. In a few cases this chromosome switched to the Cα gene [5].

The induction of double-staining cells within a few days after treatment with LPS indicates that LPS induces switching in culture. Up to 2% of the cells are double-staining 4 days after addition of LPS [5]. However, it is clear that the IgA$^+$ cells observed after LPS treatment can result from expansion of single IgA$^+$ clones. By examining the Sμ-Sα junctions in expressed α-genes cloned from cells from one experiment 23 days after initiation of the cultures, we found that the expressed α-genes were derived from one switch recombination event that occurred prior to the addition of LPS since the Sμ-Sα junctions were identical in IgA cells from two *different* wells of the same experiment [7]. Thus, although active switching does occur in these cultures as shown by the presence of double-staining cells, the actual frequency of productive switching, leading to an IgA$^+$ cell that proliferates in culture, must be low. However, we also found that the Sμ-Sα junctions from IgA$^+$ cells derived from different experiments generally do differ [3, 4, 7, 8]. We have shown that switch recombination appears to occur by a copy-choice mechanism involving error-prone DNA synthesis [7, 8]. IgA$^+$ cells increase relative to IgM$^+$ cells for about 2 weeks after treatment with LPS for 4 days, but then the proportion of cells that are IgM$^+$ begin to increase and by about 3–4 weeks the cultures are ≥99% IgM$^+$.

Lymphokines Influence Switching by I.29μ Cells

T cells are not required for induction of switching by I.29μ cells. However, in experiments we have performed in collaboration with Eva Severinson of the University of Stockholm, we have found that interleukin-4 (IL-4) and a supernatant from a T_{H2} cell line (2.19) augment LPS-induced switching by about 1.5-fold [9]. The 2.19 supernatant contains IL-3, IL-4, IL-5 and other factors [10, 11; Eva Severinson, pers. commun.]. The effects of IL-4 and the 2.19 supernatant on I.29μ cells differ somewhat from their effect on spleen cells. The first difference is that IL-4 alone does not increase switching to IgA$^+$ by LPS-activated spleen cells. This may be due to the fact that I.29μ cells appear to already have received a signal that predisposes them to switch to IgA and that IL-4 can synergize with this signal. The second difference is that the magnitude of the effect of 2.19 supernatant on spleen cells is greater than its effect on I.29μ cells. The 2.19 T_{H2} supernatant increases the expression of IgA by LPS-treated spleen cells by 5- to 20-fold [12]. This can also be explained by prior activation of the I.29μ cells, which could limit the amount of further induction possible.

IL-5 [11] treatment of I.29μ cells reduces the percentage of IgA$^+$ cells induced by LPS by about 50%, but it also causes the cells to enlarge and to look like plasma cells and it decreases cell viability [J.S., unpubl. data]. This is consistent with the report that IL-5 promotes secretion of IgA by IgA$^+$ Peyer's patch cells [13]. Interferon-γ (IFN-γ) inhibits switching to IgA and to IgE by up to 70% and this inhibition does not appear to be simply explained by reduced cell viability (see below) [P.S., unpubl. data].

Transforming growth factor-β (TGF-β) has been reported to increase the secretion of IgA by LPS-treated IgA$^-$ mouse B cells by about 10-fold, probably by actually directing the switch to IgA [14, 15]. TGF-β also increases the percentage of IgA$^+$ cells by about 2-fold when added during LPS induction of I.29μ cells [P.S., unpubl. data]. Recombinant mouse IL-3 and human IL-6 appear to have no effect on switching by I.29μ cells [P.S. and J.S., unpubl. observ.]. In experiments to be described below we have shown that IL-4, the 2.19 supernatant, IFN-γ and TGF-β act prior to the actual switch event, apparently directing or inhibiting the switch itself, rather than helping or inhibiting cells after switching.

Mechanism of Regulation of Isotype Switch

LPS induces I.29μ cells to switch mostly to IgA whereas LPS induces normal mouse spleen cells to switch mostly to IgG3 and IgG2b expression [12]. One of our first questions about the I.29 system was why this preferential switch to IgA? We have an outline of an answer to this question, which is that in the IgM$^+$ cells prior to induction of the switch the Cα gene is in an active, accessible state. The unrearranged Cα gene is hypomethylated relative to liver cell DNA (probably on both chromosomes) in I.29μ cells to at least the same extent as in IgA$^+$ cells, as assayed by using the methylation-sensitive enzyme HpaII [16]. The Cε and Cγ$_{2a}$ genes, to which the cells switch less frequently, are also hypomethylated but not nearly as completely. The Cγ$_3$ gene, to which the nonexpressed chromosome switches in vivo, is also partially hypomethylated. Genes to which the cells rarely switch, i.e. Cγ$_1$ and Cγ$_{2b}$, are fully methylated. In fact, the Cγ$_1$ gene is more methylated in I.29μ cells than it is in liver cells.

By contrast, the B cell line, BCL-1, which expresses IgM and a very small amount of IgD [17], and which has never been found to switch to IgA, contains completely methylated Cα gene fragments, indicating that not all IgM$^+$ B cell lines have hypomethylated Cα genes. The Cγ$_3$ and Cγ$_{2b}$ genes are, however, slightly hypomethylated in BCL-1 cells [16].

As genes which are transcriptionally active are generally hypomethylated, we examined whether the unrearranged Cα genes were being transcribed in IgM$^+$ cells by hybridizing blots of poly(A)$^+$ RNA from unstimulated I.29μ cells with probes for all the C$_H$ genes [16]. The IgM$^+$ cells contain small amounts of α-, ε-, γ$_{2a}$- and γ$_3$-poly(A)$^+$ RNAs which are not the size of α-, ε-, γ$_{2a}$- and γ$_3$-mRNAs, and which do not hybridize with an I.29 heavy-chain variable-region gene probe. No transcripts were detected with a γ$_1$- or γ$_{2b}$-probe. Thus, C$_H$ genes to which I.29μ cells switch are transcriptionally active.

To determine if the amount of the germ line α-RNAs correlates with the probability of switching to IgA, we examined the level of germ line α-transcripts in various clones of I.29μ cells (isolated by Roberto Sitia at the Institute for Cancer Research in Genoa, Italy) which switch to IgA to varying extents [9]. The level of germ line α-RNA present in uninduced cells of an I.29μ clone that switched the best (22A10: 9% IgA$^+$ at day 10 after LPS treatment) was about 10^2 times greater than in the clone that switched the worst (C19: 0.2% IgA$^+$). The amount of α-RNAs in other clones, which switched to intermediate extents, also correlated with their ability to switch.

The BCL-1 cell line does not contain these α- and ε-RNAs, although it does contain $γ_3$- and $γ_2$-RNAs of the same size as present in I.29μ cells, indicating that these RNAs may be initiated and processed at identical sites in I.29μ and BCL-1 cells. The fact that BCL-1 cells do not contain the germ line α- and ε-transcripts further indicates that the activation of the α-, ε- and $γ_{2a}$-genes in I.29μ cells is specifically regulated. Another cell line, the pre-B cell line 18-8, which switches specifically to IgG2b, has been shown to have germ line $γ_{2b}$-transcripts prior to switching [18].

These data support an accessibility model for regulation of specificity of switching, i.e. switching occurs preferentially to specific genes because they are in an active, or accessible, chromatin structure [16, 18, 19]. Data supporting a similar model for immunoglobulin V, D and J gene joining have been reported [20, 21]. The commitment to switch to IgA, IgE or IgG2a-may have been effected by the previous interaction of the precursor to the I.29 lymphoma with a T cell or dendritic cell or by a factor secreted by such a cell. These data argue indirectly against a model whereby isotype specificity would be determined by class-specific recombinases.

Function of Transcription of Germ Line C_H Genes

Transcription of the germ line C_H genes may occur simply because the C_H gene has been made accessible to the 'switch recombinase' by binding certain nuclear factors and as a by-product is also accessible to RNA polymerase. Alternatively, it is possible that the act of transcription promotes DNA recombination or that the RNA molecules themselves participate in the switch process, perhaps by specifying the site of recombination, or perhaps the RNAs are translated and their product directs switching. Evidence from yeast recombination [22] and Vκ-Jκ recombination in transfected plasmids [21] supports the notion that the act of transcription can promote recombination. Below we will present evidence suggesting that the germ line α-RNA may be translated. These different possibilities are not mutually exclusive.

Structure of Germ Line RNAs

To begin to understand the function of transcription of germ line C_H genes, we have determined the structure of the predominant RNA transcribed from the unrearranged α-gene in I.29μ cells. By RNA blotting, analysis of cDNAs, RNase protection and primer-extension experiments, the predominant 1.7-kilobase (kb) germ line RNA was found to be encoded by an upstream 'I' exon with two initiation sites located about 2.5 kb 5′ to the Sα sequences [23]. Figure 1 presents transcription maps of the 1.7- and 3-kb germ line α-RNAs, the latter of which has the membrane domain present in mRNA for membrane-bound α-chains [9]. In addition to the 1.7- and 3-kb germ line α-RNAs, I.29μ cells have 4 smaller germ line α-RNAs whose structure we have not determined. As true for the germ line $γ_{2b}$- and $γ_1$-transcripts, which also have multiple initiation sites [24, 25], there are no TATA or CCAAT motifs located nearby upstream. No antisense α-transcripts could be detected. Thus, transcription of the germ line α-gene proceeds through Sα sequences which could aid in opening up this region for recombination.

Germ Line Transcripts Are Induced in Spleen Cells by Interleukins Which Direct Switching to Specific Classes

To determine if germ line transcripts were induced in normal cells under conditions that induce switching to specific isotypes, we examined, in collaboration with Eva Severinson at the University of Stockholm, whether lymphokines that influence switching would induce transcripts from unrearranged C_H genes prior to switching. Specifically, we questioned whether IL-4 would induce transcripts from unrearranged $Cγ_1$ and Cε genes since it has been shown that IL-4 will cause LPS-activated spleen cells to switch to IgG1 and IgE, whereas in the absence of IL-4, LPS-activated spleen cells switch to IgG3 and IgG2b [10, 12, 26]. By hybridizing blots of poly(A)$^+$ RNA isolated from spleen cells 2 days after treatment with IL-4 with a probe to the region 5′ to the $Sγ_1$ region or with a Cε probe, we found that IL-4 alone induced germ line $Cγ_1$ transcripts and IL-4 plus LPS induced germ line Cε transcripts [9]. Induction of germ line Cε transcripts required at least 40 times more IL-4 than induction of germ line $Cγ_1$ transcripts; this is consistent with the fact that much more IL-4 is required to induce switching to IgE than to IgG1 [27]. Switching is first detected in these cultures 3 days after LPS treatment [28].

We also examined the effect of the supernatant from the T_{H2} cell line (2.19) [12], which induces LPS-activated spleen cells to switch to IgG1, IgE and IgA, on the level of germ line transcripts in spleen cells [9]. Treatment with the 2.19 supernatant for 2 days induced 1.7-kb germ line $C\alpha$ transcripts in addition to germ line $C\gamma_1$ and $C\epsilon$ transcripts. IL-4, which is present in this supernatant [10], given alone did not induce germ line $C\alpha$ transcripts, and we do not know what factor (or factors) in this T_{H2} supernatant induces germ line $C\alpha$ transcripts. These experiments demonstrated that lymphokines that promote switching to specific heavy-chain classes in the presence of LPS induce RNA transcripts from these same C_H genes before the actual switch recombination or expression of the new isotype.

Germ line transcripts of all the C_H genes have been identified except for $C\delta$. Their structures are analogous, i.e. they have an I exon transcribed from the region 5′ to switch sequences which is spliced to exons encoding the C_H domain [23–25; E. Severinson et al., submitted; P.S., unpubl. data; P. Rothman and F. Alt, pers. commun.].

Additional Lymphokines Affect the Level of α- and
ε-Transcripts in I.29μ Cells

We have tested the effects of two other lymphokines that influence switching by LPS-treated I.29μ cells: IFN-γ, which inhibits switching to IgA and IgE, and TGF-β, which increases switching to IgA [P.S., unpubl. data]. Treatment with IFN-γ (10 units/ml) in the presence of LPS for 2 days reduced the level of germ line α-transcripts by 6-fold relative to LPS alone and ε-transcripts became undetectable (fig. 2a). Treatment with IFN-γ also inhibited the induction of these transcripts by IL-4 [P.S., unpubl. data]. The inhibition was quantitated by comparison to probes for RNAs that were unaffected under these conditions, A50 [29] or GAPDH [30]. Thus, the decrease in germ line α- and ε-RNAs was not simply due to reduction in cell viability. Surprisingly, IFN-γ treatment in the absence of LPS had no effect on the level of germ line α-transcripts. It has been demonstrated that IFN-γ can inhibit the IL-4-induced secretion of IgG1 or IgE by LPS-activated spleen cells [31]. Although our data indicate that IFN-γ can reduce the level of germ line α- and ε-transcripts in the absence of IL-4, the reduction of α-transcripts was only observed in the presence of LPS. We have not examined ε-transcripts after treatment with IFN-γ alone. Thus, IFN-γ may

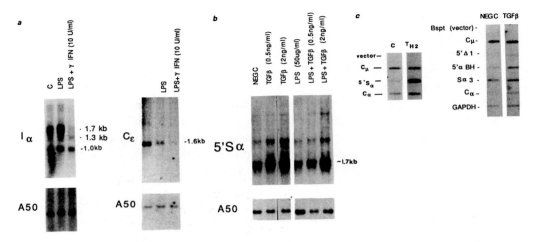

Fig. 2. RNA blots and nuclear run-on transcription assays on I.29μ cells treated with lymphokines and LPS. *a* Blots of poly(A)⁺ RNA (2 μg per lane in left panel; 1.5 μg per lane in right panel) from I.29μ cells (clone 22D) that had been treated as indicated for 2 days and hybridized with a probe for the Iα exon (BamHI/HindIII fragment) or with a probe for Cε [9]. After these probes were removed by thermal denaturation the blots were hybridized with a probe for a noninduced RNA (A50) [29]. *b* Blot of poly(A)⁺ RNA (3.1 μg per lane) from I.29μ (22D) cells treated for 2 days as indicated and hybridized with the identical probes used in *a*. *c* Run-on transcription assays on nuclei isolated from 22D cells treated for 24 h with the 2.19 supernatant (T_{H2}) or for 8 h with TGF-β (2 ng/ml). The ^{32}P-labeled RNA was hybridized with slot blots of 5 μg of plasmid DNA containing the following inserts: vector = no insert; Cμ = genomic HindIII fragment in pM2-5B [42]; 5'Δ1 = 0.7 BamHI/HhaI fragment located immediately 5' to the initiation sites of germ line α-RNA; 5' Sα or 5'αBH = Iα probe; Sα3 = 320-bp Sau3A fragment containing consensus Sα tandem repeats; Cα = genomic gene fragment encoding Cα domain [9]; GAPDH = noninducible gene [30].

exert its inhibitory effect through something induced by LPS rather than directly on the Cα gene or Cα transcripts.

Treatment of I.29μ cells with TGF-β (2 ng/ml) in the presence or absence of LPS induced a 4- to 8-fold increase in the level of germ line α-transcripts but had no effect on the level of ε-transcripts, in accord with the fact that TGF-β induces LPS-treated IgA⁻ spleen B cells to express IgA but not IgE [14, 15] (fig. 2b, and data not shown). Thus, TGF-β induces transcripts from the unrearranged α-gene and this is followed by switching to IgA in the presence of LPS. The source of TGF-β which would be physiologically relevant to the IgA switch in vivo is unknown, as a variety of cell types synthesize it [32]. It has been reported that B cell lines synthesize TGF-β [33] and so it is possible that the 2.19 supernatant that induces germ line α-transcripts and switching to IgA may be inducing synthesis or activation of TGF-β synthesized by I.29μ cells or by spleen cells.

Lymphokines Induce Transcription of the Germ Line α-Gene in I.29μ Cells

If lymphokines direct switching by inducing accessibility of specific C_H genes, one would expect they would induce transcription factors which would bind to sites in the genome that regulate transcription of germ line C_H genes. To determine if the lymphokines that have been shown by RNA blotting to induce germ line α-transcripts do so by inducing RNA transcription, we have performed nuclear run-on experiments after treatment of I.29μ cells (22D and 22A10 clones) with various lymphokines. Treatment with the 2.19 supernatant for 24 h in the absence of LPS induces transcription across the Iα exon by about 20-fold (fig. 2c). Surprisingly, transcription was not as strongly induced by treatment with the 2.19 supernatant in the presence of LPS, even at other time points (data not shown). Also surprising was the finding that transcription across the Cα region was only induced about 2-fold by the 2.19 supernatant (fig. 2c). Thus, there appears to be a large induction of transcription which is prematurely terminated. This RNA appears to be labile since we have never detected on blots an RNA species that hybridizes with the Iα but not the Cα probe. Transcription of other genes, e.g. c-myc, has also been shown to be regulated by premature termination or pausing [34, 35]. No transcription was detected using a probe for the region 5′ to the initiation sites of the germ line α-transcripts nor with a probe for antisense transcription (fig. 2c, and data not shown). Treatment with IL-4 for 18 h induced transcription across the Iα and Cα exons by about 2-fold (not shown). The 2.19 supernatant and Il-4 induce the steady-state level of germ line α-RNAs by 1.5- to 5-fold as assayed by RNA blotting [9].

TGF-β (2 ng/ml) induced transcription across both the Iα and Cα exons by 3-fold but with faster kinetics than did the 2.19 supernatant or IL-4 (fig. 2c). A 2-fold induction was seen as early as 4 h after TGF-β was added to the cells (data not shown). Neither IL-5 nor IL-6 in the presence or absence of LPS or IL-4 had any effect on transcription across the Iα exon.

In conclusion, lymphokines which induce germ line α-transcripts as assayed by RNA blotting and which augment switching to IgA by I.29μ or spleen cells induce transcription of the unrearranged α-gene prior to switch recombination. The level of germ line α-transcripts is regulated by the rate of transcription but also postranscriptionally. LPS does not induce transcription although it does increase the level of the germ line α-transcripts detected on RNA blots by 2- to 20-fold [9] (data not shown). Furthermore, the marked induction of transcription by the 2.19 supernatant across the Iα exon was not accompanied by accumulation of a poly(A)$^+$ RNA detectable by blotting. The finding that the germ line α-RNA appears to be regulated at the posttranscriptional level, in addition to the transcriptional level, suggests that the germ line α-RNAs themselves, or a product encoded by these RNAs, have a function. However, since much of the regulation by lymphokines does appear to occur at the transcriptional level, the data are still consistent with the hypothesis that the act of transcription functions to direct recombination and this could be the primary function of transcription of germ line C_H genes.

Identification of Promoter Sequences for Transcription of Germ Line α-Genes

In order to analyze how transcription of the germ line α-gene is regulated, we have examined whether DNA sequences 5′ to the initiation sites for the germ line α-RNA can promote transcription from a reporter gene after transfection into I.29μ cells [Y.C.L., unpubl. data]. We found that the 5′ flanking sequences do function to promote expression of the gene for chloramphenicol acetyl transferase (CAT) after transfection into I.29μ cells (clone 22D) but not after transfection into several cell lines that do not express germ line transcripts (including a myeloma and a B cell hybridoma, a T cell line and 2 pre-B cell lines) [Y.C.L. and A. De Pass, unpubl. data].

At least some of the RNAs transcribed from these constructs initiated at the correct sites. Maximal expression of the CAT gene was obtained if about 500 nucleotides 5′ to the initiation sites was included in the plasmid, and if less than this was included the expression was reduced. In conclusion, the region 5′ to the initiation sites for the germ line α-RNA does have promoter activity in the appropriate cell types. We are in the process of attempting to define the sequences required for regulated transcription of the germ line α-gene.

Location of Germ Line α-RNA on Polysomes

When the Iα exon was sequenced, we noticed a small open-reading frame (ORF) which would be initiated by the most 5' AUG codon in this exon and would encode a 43-amino-acid polypeptide whose termination site would be within this exon [23]. The nucleotide sequence surrounding the initiator AUG codon should form a relatively efficient translation start site since it has purines at the −3 and +4 positions [36, 37]. To determine if the germ line α-RNA may be translated in I.29μ cells, we looked for its presence on polysomes in untreated cells or in cells treated with LPS for 2 days [23]. The major 1.7-kb germ line α-RNA was found to co-sediment with small polysomes containing 1, 2 or 3 ribosomes, consistent with the small size of the ORF. In the absence of LPS about one half of the germ line RNA co-sedimented with polysomes and the remainder sedimented at the top of the gradient, whereas in the presence of LPS all the 1.7-kb α-RNA co-sedimented with small polysomes. These data suggest that the 1.7-kb germ line α-RNA may be translated and that the translation may be induced by LPS treatment. The predicted amino acid sequence of the ORF is not homologous to any protein in the GenBank database and does not have any motifs that allow us to predict its cellular location or function. Germ line α-RNA consists of about 0.06% of the poly(A)$^+$ RNA in I.29μ cells; this is abundant enough to suggest that it may have a function. The germ line γ_1-RNA also has an ORF, of 48 amino acids, with a good initiator AUG, although we have not determined whether it co-sediments with polysomes [25]. However, not all germ line RNAs may be capable of being translated. The germ line γ_{2b}-RNA has an ORF of 48 amino acids, but this ORF begins at the third AUG of the exon and has a poor context for initiation of translation [24]. Thus, it is possible that not all germ line C_H RNAs are translated and that their main function may not involve the putative translation products. We are in the process of attempting to detect the putative Iα protein in I.29μ cells.

A Nuclear Protein That Binds near Switch Regions

To attempt to identify nuclear proteins that may be involved in switch recombination, we have used the electrophoretic mobility shift assay to search for B-cell-specific proteins that bind to or near the tandemly repeated Sα sequences [38]. We have identified one B-cell-specific nuclear protein (Sα-BP) that binds to two DNA sites located about 0.4 and 1 kb 5' to the Sα tandem repeats (fig. 1). These two sites are located about 2 kb 3′ to the initiation sites of the major germ line α-RNAs. Sα-BP also binds to a site about 0.3 kb 5′ to the initiation sites of the germ line γ_1-transcripts and near the 5′ end of the Sμ region [K.U.S., unpubl. data]. We have not yet determined whether it binds near other switch sequences. SαBP has tissue- and differentiation-stage-specific expression, as it is expressed in B and pre-B cell lines but not in myeloma cell lines, nor in 2 T cell lines, nor in a fibroblast, a glioma or a pheochromocytoma cell line [38]. It is expressed in normal spleen cells and its expression does not appear to be further inducible by LPS or by IL-4. SαBP is expressed in the I.29μ line but it is also expressed in B cell lines that do not undergo switch recombination, e.g. IgA+ and IgE+ cell lines derived from the I.29 lymphoma and in the BCL-1 line. Thus, its presence does not appear sufficient to enable a cell to undergo switch recombination, although it may be necessary for switch recombination. It is absent from a hybridoma cell line derived from I.29. It has been shown that occasionally myelomas and B cell hybridomas can switch antibody class in culture by a DNA recombination event, but these DNA recombinations never involve the S regions and generally utilize sequences quite far away from the S regions [39]. Thus, Sα-BP may be required to undergo switch recombination utilizing S sequences. Atempts to determine the function of this protein have failed thus far. Multimers of the binding sites did not enhance expression from a heterologous promoter, that for *c-fos*, but we have not yet tested the homologous germ line α-promoter. Other possible functions for Sα-BP include localization of S sequences to nuclear sites where recombination occurs, or to serve to initiate the DNA synthesis that occurs during switch recombination [7, 8].

Future Directions

We have provided strong correlative evidence that transcription of germ line C_H genes serves to direct switch recombination. One of our immediate goals is to attempt to determine by direct experiments whether the process of transcription across S regions promotes switch recombination. It is also important to learn how transcription from the germ line C_H genes is regulated, i.e. to determine which lymphokines and nuclear factors induce or repress transcription and which DNA sequences are necessary for this regulation. Understanding the regulation of transcription and posttranscriptional regulation of the germ line C_H genes will help us understand how switch recombination is directed. Although lymphokines induce transcription of specific C_H genes, LPS treatment is required to obtain switching. We hypothesize that LPS induces the switch recombinase and perhaps other functions required for switching. It appears to induce association of the germ line α-RNA with polysomes. Thus, it will be interesting to investigate the role of LPS by characterizing the molecules that it induces.

Acknowledgments

We thank Dr. Eva Severinson for 2.19 supernatant and for recombinant IL-4; Dr. Fritz Melchers of the Basel Institute for Immunology, Switzerland, for the IL-3-producing 3Ag8–653 cell line [40] and Dr. Dana Fowlkes of the Department of Pathology, University of North Carolina, for purified human recombinant IL-6 [41]. The Schering Corporation provided the mouse recombinant IFN-γ as part of the American Cancer Society's program on interferon.

References

1 Sato H, Boyse EA, Aoki T, Iritani C, Old LJ: Leukemia-associated transplantation antigens related to murine leukemia virus. The X.1 system: Immune response controlled by a locus linked to H-2. J Exp Med 1973;138:593–606.
2 Sitia R, Rubartelli A, Hammerling U: Expression of two immunoglobulin isotypes, IgM and IgA, with identical idiotype in the B cell lymphoma I.29. J Immunol 1981; 127:1388–1394.
3 Stavnezer J, Marcu KB, Sirlin S, Alhadeff B, Hammerling U: Rearrangements and deletions of immunoglobulin heavy chain genes in the double-producing B cell lymphoma I.29. Mol Cell Biol 1982;2:1002–1013.
4 Klein D, Blance J, Sirlin S, Stavnezer J: I.29 B lymphoma cells express a non-mutated immunoglobulin heavy chain variable region gene before and after heavy chain switch. J Immunol 1988;140:1676–1684.
5 Stavnezer J, Sirlin S, Abbot J: Induction of immunoglobulin isotype switching in cultured I.29 B lymphoma cells: Characterization of the accompanying rearrangements of heavy chain genes. J Exp Med 1985;161:577–601.
6 Tada N, Kimura S, Binari R, Liu Y, Hammerling U: New mouse immunoglobulin A heavy chain allotype specificities detected using the hybridoma-derived IgA of I/St mice. Immunogenetics 1981;13:475–481.
7 Dunnick W, Wilson M, Stavnezer J: Mutations, duplication, and deletion of recombined switch regions suggest a role for replication in the immunoglobulin heavy-chain switch. Mol Cell Biol 1989;9:1850–1856.
8 Dunnick W, Stavnezer J: Copy-choice mechanism of immunoglobulin heavy chain switch recombination. Mol Cell Biol 1990;10:397–400.
9 Stavnezer J, Radcliffe G, Lin Y-C, Nietupski J, Berggren L, Sitia R, Severinson E: Immunoglobulin heavy-chain switching may be directed by prior induction of transcripts from constant region genes. Proc Natl Acad Sci USA 1988;85:7704–7708.

10 Noma Y, Sideras P, Naito T, Bergstedt-Lindqvist S, Azuma C, Severinson E, Tanabe
    T, Kinashi T, Matsuda F, Yaoita Y, Honjo T: Cloning of cDNA encoding the murine
    IgG1 induction factor by a novel strategy using SP6 promoter. Nature 1986;319:
    640–646.
11 Kinashi T, Harada N, Severinson E, Tanabe T, Sideras P, Konishi N, Azuma C,
    Tominaga A, Bergstedt-Lindqvist S, Takahashi M, Matsuda F, Yaoita Y, Takatsu K,
    Honjo T: Cloning of complementary DNA encoding T-cell replacing factor and
    identity with B-cell growth factor II. Nature 1986;324:70–73.
12 Sideras P, Bergstedt-Lindqvist P, MacDonald HR, Severinson E: Secretion of IgG1
    induction factor by T cell clones and hybridomas. Eur J Immunol 1985;15:586–593.
13 Harriman GR, Kunimoto DY, Elliott JF, Paetkau V, Strober W: The role of IL-5 in
    IgA B cell differentiation. J Immunol 1988;140:3033–3039.
14 Coffman RL, Lebman DH, Shrader B: Transforming growth factor β specifically
    enhances IgA production by lipopolysaccharide-stimulated murine B lymphocytes.
    J Exp Med 1989;170:1039–1044.
15 Sonada E, Matsumoto R, Hitoshi Y, Ishii T, Sugimoto M, Araki S, Tominaga A,
    Yamaguchi N, Takatsu K: Transforming growth factor β induces IgA production and
    acts additively with interleukin 5 for IgA production. J Exp Med 1989;170:1415–
    1420.
16 Stavnezer-Nordgren J, Sirlin S: Specificity of immunoglobulin heavy chain switch
    correlates with activity of germline heavy chain genes prior to switching. EMBO J
    1986;5:95–102.
17 Severinson-Gronowicz E, Doss C, Howard D, Morrison DC, Strober S: An in vitro
    line of the B cell tumor BCL$_1$ can be activated by LPS to secrete IgM. J Immunol
    1980;125:976–980.
18 Yancopoulos GD, Alt FW: Secondary genomic rearrangement events in pre-B cells:
    V$_H$DJ$_H$ replacement by a LINE-1 sequence and directed class switching. EMBO J
    1986;5:3259–3266.
19 Stavnezer J, Abbot J, Sirlin S: Immunoglobulin heavy chain switching in cultured
    I.29 murine B lymphoma cells: Commitment to an IgA or IgE switch. Curr Top
    Microbiol Immunol 1984;113:109–116.
20 Yancopoulos GD, Alt FW: Developmentally controlled and tissue-specific expres-
    sion of unrearranged V$_H$ gene segments. Cell 1985;40:271–281.
21 Blackwell TK, Moore MW, Yancopoulos GD, Suh H, Lutzker S, Selsing E, Alt FW:
    Recombination between immunoglobulin variable region gene segments is enhanced
    by transcription. Nature 1986;324:585–589.
22 Thomas BJ, Rothstein R: Elevated recombination rates in transcriptionally active
    DNA. Cell 1989;56:619–630.
23 Radcliffe G, Lin Y-C, Julius M, Marcu KB, Stavnezer J: Structure of germline
    immunoglobulin α heavy-chain RNA and its location on polysomes. Mol Cell Biol
    1990;10:382–386.
24 Lutzker S, Alt FW: Structure and expression of germline immunoglobulin γ$_{2b}$ tran-
    scripts. Mol Cell Biol 1988;8:4585–4588.
25 Xu M, Stavnezer J: Structure of germline immunoglobulin heavy chain γ$_1$ tran-
    scripts. Dev Immunol 1990;1: in press.
26 Coffman RL, Ohara J, Bond MW, Carty J, Zlotnik A, Paul WE: B cell stimulatory
    factor-1 enhances the IgE response of lipopolysaccharide-activated B cells. J Im-
    munol 1986;136:4538–4541.

27 Snapper CM, Finkelman FD, Paul WE: Differential regulation of IgG$_1$ and IgE synthesis by interleukin 4. J Exp Med 1988;167:183–196.

28 Winter E, Krawinkel U, Radbruch A: Directed Ig class switch recombination in activated murine B cells. EMBO J 1987;6:1663–1671.

29 Nguyen HG, Medfod RM, Nadal-Ginard B: Reversibility of muscle differentiation in the absence of commitment: Analysis of a myogenic cell line temperature-sensitive for commitment. Cell 1983;34:281–293.

30 Fort P, Marty L, Piechaczyk M, El Sabrouty S, Dani C, Jeanteur P, Blanchard JM: Various rat adult tissues express only one major mRNA species from the glyceralde-hyde-3-phosphate dehydrogenase multigenic family. Nucleic Acids Res 1985;13: 1431–1442.

31 Coffman RL, Carty J: A T cell activity that enhances polyclonal IgE production and its inhibition by interferon-γ. J Immunol 1986;136:949–954.

32 Wahl SM, McCartney-Francis N, Mergenhagen SE: Inflammatory and immuno-modulatory roles of TGF$_\beta$. Immunol Today 1989;8:258–261.

33 O'Garra A, Barbis D, Whitmore AC, Haughton G, Pearce MK, Lee F, Arnold LW, Dhar V, Stapleton G, Rennick D, Howard M: Constitutive production of known and novel lymphokines by murine B lymphomas. Int Congr Immunol Abstr 1989;7:378.

34 Bentley DL, Groudine M: A block to elongation is largely responsible for decreased transcription of c-myc in differentiated HL-60 cells. Nature 1986;321:702–706.

35 Nepveu A, Marcu KB: Intragenic pausing and anti-sense transcription within the murine c-myc locus. EMBO J 1986;5:2859–2865.

36 Kozak M: Point mutations define a sequence flanking the AUG initiator codon that modulates translation by eukaryotic ribosomes. Cell 1986;44:283–292.

37 Kozak M: At least six nucleotides preceding the AUG initiator codon enhance translation in mammalian cells. J Mol Biol 1987;196:947–950.

38 Waters SH, Saikh KU, Stavnezer J: A B-cell specific nuclear protein that binds to DNA sites 5′ to immunogobulin Sα tandem repeats is regulated during differentia-tion. Mol Cell Biol 1989;9:5594–5601.

39 Sablitsky F, Radbruch A, Rajewsky K: Spontaneous immunoglobulin class switching in myeloma and hybridoma cell lines differs from physiological class switching. Immunol Rev 1982;67:59–73.

40 Karasugama H, Melchers F: Establishment of mouse cell lines which constitutively secrete large quantities of interleukin 2, 3, 4 or 5 using modified cDNA expression vectors. Eur J Immunol 1988;18:98–104.

41 Jambou RC, Snouwaert JN, Bishop GA, Stebbins JR, Frelinger JA, Fowlkes DM: High level expression of a bioengineered, cysteine-free hepatocyte-stimulating factor (interleukin 6)-like protein. Proc Natl Acad Sci USA 1988;85:9426–9430.

42 Marcu KB, Banerji J, Penncavage NA, Lang R, Arnhein N: 5′ Flanking region of immunoglobulin heavy chain constant region displays length heterogeneity in germ-lines of inbred mouse strains. Cell 1980;22:187–196.

Janet Stavnezer, MD, Department of Molecular Genetics and Microbiology,
University of Massachusetts Medical School, Worcester, MA 01655 (USA)

Subject Index